United States Nuclear Regulatory Commission

Protecting People and the Environment

A Short History of Nuclear Regulation, 1946–2009

by J. Samuel Walker and Thomas R. Wellock

History Staff
Office of the Secretary
U.S. Nuclear Regulatory Commission

I0482681

October 2010

Preface

"History," automobile maker Henry Ford once said, "is more or less...bunk." Philosopher George Santayana was more charitable in his assessment of this discipline when he declared that "those who fail to study the past are condemned to repeat it." In a sense, both Ford and Santayana were right. Much of the past has little meaning or importance for the present and deservedly remains forgotten in the dustbins of history. However, other parts of the past need to be remembered and studied in order for us to make sense from the present. Today's events are a direct outgrowth of yesterday's events, and understanding the history of any given problem is essential to approaching it knowledgeably. It is the task of the historian to gather evidence, to separate what is important from what is not, and to explain key events and decisions of the past.

This short history of nuclear regulation provides a brief overview of the most significant events in the U.S. Nuclear Regulatory Commission's past. Space limitations prevent discussion of all the important occurrences, and even the subjects that are included cannot be covered in full detail. The first chapter of this account is taken from George T. Mazuzan and J. Samuel Walker, Controlling the Atom: The Beginnings of Nuclear Regulation, 1946–1962 (University of California Press, Berkeley, CA, 1984). The second chapter is largely based on J. Samuel Walker, Containing the Atom: Nuclear Regulation in a Changing Environment, 1963–1971 (University of California Press, Berkeley, CA, 1992). The third chapter is adopted in significant part from J. Samuel Walker, Three Mile Island: A Nuclear Crisis in Historical Perspective (University of California Press, Berkeley, CA, 2004). The findings and conclusions on events that occurred after 1979 should be regarded as preliminary and tentative; they are not based on extensive research in primary sources. However, we hope that this overview will help explain how the past has shaped the present and will illuminate the considerations that have influenced regulatory decisions and procedures over the years. We also hope that this outline will suggest that history should be viewed as something more valuable than mere "bunk."

Contents

Chapter One

The Formative Years of Nuclear Regulation, 1946–1962

Chapter 1

The use of atomic bombs against the Japanese cities of Hiroshima and Nagasaki in August 1945 ushered in a new historical epoch, breathlessly labeled in countless news reports, magazine articles, films, and radio broadcasts as the "atomic age." Within a short time after the end of World War II, politicians, journalists, scientists, and business leaders suggested that peaceful applications of nuclear power could be as dramatic in their benefits as nuclear weapons were awesome in their destructive power. Nuclear physicist Alvin M. Weinberg told the U.S. Senate's Special Committee on Atomic Energy in December 1945, "Atomic power can cure as well as kill. It can fertilize and enrich a region as well as devastate it. It can widen man's horizons as well as force him back into the cave." Newsweek reported that "even the most conservative scientists and industrialists [are] willing to outline a civilization which would make the comic-strip prophecies of Buck Rogers look obsolete." Observing that ideas for the civilian uses of atomic energy ranged "from the practical to the fantastic," it cited a few examples: (1) atomic-powered airplanes, rockets, and automobiles, (2) large electrical generating stations, (3) small "home power plants" to provide heat and electricity to individual homes, and (4) tiny atomic generators wired to clothing to keep a person cool in summer and warm in winter.

Developing nuclear energy for civilian purposes, as even the most enthusiastic proponents recognized, would take many years. The Government's first priority was to maintain strict control over atomic technology and to investigate its military applications. The Atomic Energy Act of 1946, which was passed as tensions with the Union of Soviet Socialist Republics (U.S.S.R.) were developing into the cold war, acknowledged, in passing, the potential peaceful benefits of atomic power. However, it emphasized the military aspects of nuclear energy and underscored the need for secrecy and the continued production of weapons. The 1946 law did not allow for private, commercial application of atomic energy; instead, it created a virtual Government monopoly of the technology. To manage the Nation's atomic energy programs, the Atomic Energy Act of 1946 established the five-member U.S. Atomic Energy Commission (AEC).

The Atomic Energy Act of 1954

By 1954, the same Cold War calculations that had earlier curtailed the commercial uses of atomic energy led Federal officials to reverse course. The initial impetus for peaceful atomic development came mostly from considerations other than meeting America's energy demands. In the early 1950s, projections of future energy requirements predicted that atomic power would eventually play an important role in the Nation's energy supplies, but these projections did not suggest an immediate need for the construction of atomic power reactors. The prevailing sense of urgency, at least among Government leaders, reflected instead the fear of falling behind other nations in fostering peaceful atomic progress. The strides that Great Britain was making in the field seemed disturbing enough, but the possibility that the U.S.S.R. might surpass the United States in civilian power development was even more ominous. AEC Commissioner Thomas E. Murray described a "nuclear power race" in a 1953 speech and warned that the "stakes are high." He added, "Once we become fully conscious of the possibility that power hungry countries will gravitate toward the U.S.S.R. if it wins the nuclear power race…it will be quite clear that this power race is no Everest-climbing, kudos-providing contest." Like Murray, many Government officials emphasized that surrendering America's lead in expanding the peaceful applications of atomic energy would deal a severe blow to its international prestige and world scientific dominance.

The eagerness to push for rapid civilian nuclear development was intensified by an impulse to show that atomic technology could serve both constructive and destructive

Chapter 1

purposes. The assertions made shortly after World War II that atomic energy could provide spectacular advances that would raise living standards throughout the world remained unproven and largely untested. As the nuclear arms race took on more terrifying proportions with the development of thermonuclear bombs, the desire to demonstrate the benefits of atomic energy became more acute. President Dwight D. Eisenhower, spurred by the detonation of the U.S.S.R.'s first hydrogen device, starkly depicted the horror of nuclear warfare in a widely publicized address to the United Nations in December 1953. At the same time, he emphasized that "this greatest of all destructive forces can be developed into a great boon, for the benefit of all mankind." Many other high Government officials echoed Eisenhower's appeal for peaceful nuclear progress and his affirmation of the potential blessings of civilian atomic energy.

By 1954, a broad political consensus viewed the development of nuclear energy for civilian purposes as a vital goal. In that year, Congress passed a new Atomic Energy Act that resulted partly from perceptions of a long-range need for new energy sources but mostly from the immediate commitment to maintain America's world leadership in nuclear technology, enhance its international prestige, and demonstrate the benefits of peaceful atomic energy. Those considerations infused the atomic power program with a sense of urgency. The Atomic Energy Act of 1954, as amended, permitted for the first time the broad use of atomic energy for peaceful applications. It redefined the atomic energy program by ending the Government's monopoly on technical data and making the growth of a commercial nuclear industry an important national goal. The act directed the AEC to "encourage widespread participation in the development and utilization of atomic energy for peaceful purposes."

The Atomic Energy Act of 1954 also instructed the AEC to prepare regulations that would protect public health and safety from radiation hazards. Thus, it assigned the agency three major functions: (1) to continue its weapons program, (2) to promote the commercial uses of nuclear power, and (3) to protect against the hazards of those peaceful applications. Those functions were in many ways inseparable and proved to be incompatible when they were carried out by a single agency. The competing responsibilities and the precedence that the AEC gave to its military and promotional duties gradually damaged its credibility on regulatory issues and undermined public confidence in its safety programs.

The Atomic Energy Commission and the Development of Commercial Nuclear Power

The Atomic Energy Act of 1954 gave the AEC wide discretion on how to proceed in establishing its promotional and regulatory policies. Despite the general agreement on ultimate objectives, the means by which these objectives should be accomplished soon created sharp philosophical differences between the AEC and its congressional oversight committee, the Joint Committee on Atomic Energy. The AEC favored a partnership between Government and industry in which private firms would play an integral role in demonstrating and expanding the use of atomic power. "The Commission's program," AEC Chairman Lewis L. Strauss explained, "is directed toward encouraging development of the uses of atomic energy in the framework of the American free enterprise system." He added that it was the AEC's conviction "that competitive economic nuclear power…would be most quickly achieved by construction and operation of fullscale plants by industry itself." To accomplish its objectives, the AEC announced a "power demonstration reactor program" in January 1955. The agency offered to perform research and development on power reactors in its national laboratories, to subsidize additional research undertaken by industry through fixed-sum contracts, and to waive for 7 years the fuel-use charges for the loan of fissionable materials that the Government would continue to own. For their part, private utilities and vendors would supply the capital for the construction of nuclear plants and pay operating expenses other than fuel charges. The purpose of the demonstration program was to stimulate private participation and investment in exploring the technical and economic feasibility of different reactor designs. At that time, no single reactor type had clearly emerged as the most promising of the several that had been proposed.

The AEC also sought to meet industry demands for technical information. For several years, some utility executives had shown a keen interest in investigating the use of nuclear fission for generating electricity. However, commercial applications of atomic energy had been thwarted by the severe limitations placed on access to information as dictated by the Atomic Energy Act of 1946. In 1953, when the Joint Committee conducted public hearings on peaceful atomic development, spokesmen for private firms emphasized that industrial progress was possible only if the restrictions on obtaining data were eased. By opening nuclear technology to commercial applications, the Atomic Energy Act of 1954 largely satisfied those complaints. From the utility companies' perspective, the Atomic Energy Act of 1954 offered companies an opportunity to participate in nuclear development and gain experience in a technology that promised to help meet long-term energy demands. Vendors of reactor components welcomed the prospects of expanding their markets not only in the United States but also in foreign countries where the need for new sources of power was more immediate.

Despite those incentives, the AEC's initiatives received a mixed response. The enthusiasm of the private utility industry for nuclear power development was tempered by several considerations. Although experiments with AEC-owned reactors had established the technical feasibility of using nuclear fission to produce electricity, many scientific and engineering questions remained unanswered. Further, the financial inducements that the AEC offered through its power demonstration reactor program did not eliminate the risks to a company's balance sheets. The capital and operating costs of atomic power were certain to be much higher than those for fossil fuel plants. Across the industry, the prospects of realizing short-term profits from nuclear power were unlikely. An American Management Association symposium in 1957 concluded, "The atomic industry has not been—and is not likely to be for a decade—attractive as far as quick profits are concerned." When Lewis Strauss made his oft-quoted statement in 1954 that nuclear power could provide electricity "too cheap to meter," he was indulging in a flight of fancy. His remark did not represent the views of the AEC or the fledging nuclear industry that knew that the heavy investments required were a major impediment to the growth of nuclear power.

In addition to technical and financial considerations, recognition of the hazards of the technology intensified industry's reservations about nuclear power. Based on experience with Government test reactors and the prevailing faith in the ability of scientists and engineers to solve technological problems, the AEC and industry leaders regarded the chances of a disastrous atomic accident as remote. However, they did not dismiss the possibility entirely. Francis K. McCune, General Manager of the Atomic Products Division of General Electric, told the Joint Committee in 1954 that "no matter how careful anyone in the atomic energy business may try to be, it is possible that accidents may occur."

Mindful of both the costs and the risks of atomic power, the electric utility industry responded to the Atomic Energy Act of 1954 and the AEC's demonstration program with restraint. Although many utilities were interested in exploring the potential of nuclear power, few were willing to press ahead rapidly in the face of existing uncertainties. The AEC was gratified and rather surprised that by August 1955, five power companies—either as individual utilities or as consortiums—had announced plans to build nuclear plants. Two of these companies decided to proceed without Government assistance, and the other three submitted proposals for projects under the AEC's power demonstration program.

The Joint Committee was less impressed with the response of private industry to the Atomic Energy Act of 1954 and the AEC's incentives. The Democratic majority of the committee favored a larger Government role in accelerating nuclear development, which conflicted with the AEC's commitment to encourage maximum private participation. The issue became a major source of contention between the AEC and the Joint Committee, thus adding a philosophical dispute to the already strained political differences resulting from the bitter personal feud between Strauss and Joint Committee Chairman Clinton P. Anderson.

In 1956, two Democratic members of the Joint Committee, Representative Chet Holifield and Senator Albert Gore, introduced legislation directing the AEC to construct six pilot nuclear plants, each with a different design, to "advance the art of generation of electrical energy from nuclear energy at the maximum possible rate." Supporters of the bill contended that the United States was falling behind Great Britain and the U.S.S.R. in the quest for practical and economical nuclear power. Opponents of the measure denied that the United States had surrendered its lead in atomic technology and insisted that private industry was best able to expedite further development. Strauss declared that "we have a civilian program that is presently accomplishing far more than we had reason to expect in 1954." The Gore-Holifield bill was defeated by a narrow margin in Congress, but the views that it embodied and the Joint Committee's impatience for rapid development of atomic power placed a great deal of pressure on the AEC to show that its reactor programs were producing results.

The Atomic Energy Commission's Regulatory Program

The AEC's determination to push nuclear development through a partnership with private industry had a major impact on the agency's regulatory policies. The AEC's fundamental objective in drafting regulations was to ensure that public health and safety were protected without imposing overly burdensome requirements that would impede industrial growth. In 1955, Commissioner Willard F. Libby articulated an opinion common among AEC officials when he remarked, "Our great hazard is that this great benefit to mankind will be killed aborning by unnecessary regulation." Other proponents of nuclear development shared this view. They realized that safety was indispensable to progress; an accident could destroy the industry or at least set it back many years. At the same time, they worried that regulations that were too restrictive or inflexible would discourage private participation and investment in nuclear technology.

The inherent difficulty that the AEC faced in distinguishing between essential and excessive regulations was compounded by technical uncertainties and by limited operating experience with power reactors. The safety record of the AEC's own experimental reactors engendered confidence that safety problems could be resolved and the possibility of accidents could be kept to "an acceptable calculated risk." However, experience at that time offered little definitive guidance on some important technical and safety questions, such as the effect of radiation on the properties of reactor materials; the durability of steel and other metals under stress in a reactor; the ways in which water reacted with uranium, thorium, aluminum, and other elements in a reactor;

and the measures needed to minimize radiation exposure in the event of a large accident.

The AEC's regulatory staff, which was created soon after the passage of the Atomic Energy Act of 1954, confronted the task of writing regulations and devising licensing procedures rigorous enough to ensure safety but flexible enough to allow for new findings and rapid changes in atomic technology. Within a short period of time, the staff drafted rules and definitions on radiation protection standards, the distribution and safeguarding of fissionable materials, and the qualifications of reactor operators. It also established procedures for licensing privately owned reactors. The Atomic Energy Act of 1954 outlined a two-step procedure for granting licenses. The AEC would issue a construction permit if it found the safety analysis submitted by a utility for a proposed reactor to be acceptable. After the utility completed the construction and the AEC determined that the plant fully met safety requirements, the applicant would receive a license to load fuel and begin operation.

Because of the uncertainties in technical knowledge and the AEC's goal of encouraging different reactor designs, the agency had to judge license applications on a case-by-case basis. The early state of the technology precluded the possibility of formulating universal standards for all aspects of reactor engineering. The regulatory staff reviewed the information that applicants supplied on the suitability of the proposed site, construction specifications, a detailed plan of operation, and safety features. The proposal received further scrutiny from a panel of outside experts, the Advisory Committee on Reactor Safeguards (ACRS), which comprised part-time consultants who were recognized authorities on various aspects of reactor technology. ARCS conducted its own independent review of the application, and its recommendations and those of the staff went to the AEC Commissioners, who then made the final decision on whether or not to approve a construction permit or operating license. (Later, the Commission delegated the consideration of regulatory staff and ACRS judgments to panels drawn from the Atomic Safety and Licensing Board while retaining final jurisdiction in licensing cases if it chose to review a panel ruling.)

The AEC did not require a prospective power reactor owner to submit finalized technical data on the safety of a facility to receive a construction permit. The agency was willing to grant a conditional permit as long as the application provided "reasonable assurance" that the projected plant could be constructed and operated at the proposed site "without undue risk to the health and safety of the public." This two-step licensing system enabled the AEC to authorize the construction of nuclear plants while allowing it enough time to investigate outstanding safety questions and to prescribe modifications to initial plans. Agency officials recognized that the wisdom of permitting construction to proceed without first resolving all potential safety problems was disputable, but they saw no alternatives in light of the existing state of the technology and the commitment to the rapid development of atomic power. They were confident that regulatory requirements were adequate to guard against the hazards of nuclear generating systems. However, the AEC acknowledged that it could not eliminate all risks. ACRS Chairman C. Rogers McCullough informed the Joint Committee in 1956 that because of technical uncertainties and limited operating experience, "the determination that the hazard is acceptably low is a matter of competent judgment."

The Power Reactor Development Company Controversy

It soon became apparent that the AEC's judgment on safety issues could be influenced by its ambition to promote the private development of nuclear power. The Commission's actions in granting a construction permit for a commercial

Chapter 1

PRDC reactor under construction, 1958

fast breeder reactor, despite the reservations of ACRS, ignited an acrimonious controversy with the Joint Committee
and raised questions about the AEC's regulatory program.
In January 1956, the Power Reactor Development Company
(PRDC), a consortium of utilities led by Detroit Edison
Company, applied for a permit to build a fast breeder reactor
in Lagoona Beach, MI, located on Lake Erie within 30 miles
of both Detroit, MI, and Toledo, OH. The AEC had already
received applications for two privately financed reactors, but
the PRDC proposal was the first to come in under the power
demonstration program.

The fast breeder reactor that PRDC planned was far more
advanced in its technological complexity than the light-water models were. Scientists and engineers had greater experience and familiarity with the light-water models proposed in
earlier applications. After review of PRDC's application and
discussions with company representatives, ACRS concluded
in an internal report to the Commission that "there is insufficient information available at this time to give assurance

that the PRDC reactor can be operated at this site without public hazard." ACRS also expressed uncertainty that its questions about the reactor's safety could be resolved within PRDC's proposed schedule for obtaining an operating license. ACRS urged that the AEC expand its experimental programs with fast breeder reactors to seek more complete data on the issues raised in the PRDC application.

The public dispute over the PRDC case was triggered by statements from Chairman Strauss and Commissioner Murray in congressional budget hearings. After the AEC requested a supplemental appropriation for the civilian power program, House Appropriations Committee Chairman Clarence Cannon subjected the Commissioners to sharp criticism when they testified in June 1956 on the need for the expenditures. Cannon, a strong public power advocate, badgered Strauss about private industry's lack of progress in atomic development and suggested that PRDC had no "intention of building this reactor at any time in the determinable future." Strauss, who was anxious to show that industry was making good headway, replied, "They [PRDC] have already spent eight million dollars of their own money to date on this project. I told you they were breaking ground on August 8. I have been invited to attend the ceremony; I intend to do so." Inadvertently, he had revealed that he planned to attend the groundbreaking ceremony for a reactor whose construction permit was still being evaluated by the AEC.

During hearings the following day, Commissioner Murray, in an effort to demonstrate the need for research and development funds, disclosed the conclusions of ACRS on the PRDC application. Murray was so uneasy about the safety implications of the ACRS report that he met with Joint Committee Chairman Anderson to outline its contents. Members of the Joint Committee were angered and disturbed by Strauss' and Murray's revelations, not only because of safety concerns but also because of the AEC's fail-

ure to inform them officially about the ACRS reservations.
The AEC was obliged by the Atomic Energy Act of 1954
to keep the Joint Committee "fully and currently informed"
about its activities, and Joint Committee members believed
that, in the case of the ACRS report, the agency had failed
to carry out its charge. The Joint Committee immediately requested a copy of the ACRS document. The AEC was reluctant to agree and, after long deliberation, offered to deliver a
copy only if the Joint Committee would keep it "administra-
tively confidential." The Joint Committee refused to accept
the report under those conditions. The AEC was even less
accommodating with the State of Michigan. When Governor
G. Mennen Williams, who had learned of the ACRS report
from Senator Anderson, asked the AEC for a copy, it refused
on the grounds that "it would be inappropriate to disclose
the contents of internal documents."

Meanwhile, the AEC's regulatory staff was completing its
review of PRDC's application. The staff took a more optimistic view of the safety of the proposed reactor than ACRS
had. Because the company had agreed to perform tests to
answer the questions raised by ACRS, the staff recommended that it be granted a construction permit. On August 2,
1956, the Commission decided to issue the permit by a vote
of three to one (Murray was the dissenter). It acknowledged
the ACRS concerns by inserting the word "conditional"
in the construction permit to emphasize that the company
would have to resolve the uncertainties about safety before
it could receive an operating license. Commissioner Harold
S. Vance summarized the majority's reasoning during the
discussion of the application. "We are doing something that
we ordinarily would not do," he said, "in that we would
not ordinarily issue a construction permit unless we were
satisfied that reasonable safety requirements had been met."
However, he added, "It may be some time before reasonable
assurance can be obtained. If we were to delay the construction permit until then, it might delay a very important

program. If we didn't think that the chances were very good that all these questions would be resolved, we would not issue the permit."

The AEC's decision elicited angry protests from the Joint Committee. Congressman Holifield, citing Strauss's earlier announcement of his plans to attend the groundbreaking ceremonies for the plant, charged that the AEC Chairman was acting in a "reckless and arrogant manner." Anderson accused the agency of conducting "star chamber" proceedings and pledged that the Joint Committee would "ascertain the full facts involved in this precipitate action."

The Joint Committee soon acted to prevent a recurrence of the AEC's conduct in the PRDC case. Anderson ordered the Joint Committee staff to prepare a study of the AEC's licensing procedures and regulatory organization and to consider, as part of the study, whether separate agencies should carry out regulatory and promotional responsibilities. The staff concluded that the creation of separate agencies was inadvisable at the time, principally because of the difficulty of recruiting qualified personnel for purely regulatory functions. It did, however, suggest other reforms in the AEC's regulatory structure and procedures. Anderson implemented his staff's proposals by introducing legislation to establish ACRS as a statutory body, direct that its reports on licensing cases be made public, and require public hearings on all reactor applications. The AEC opposed all three measures but muted its objections because Anderson presented them as amendments to a bill to provide indemnity insurance for reactor owners, which the agency strongly favored.

The Price-Anderson Act

The AEC regarded indemnity legislation as essential for stimulating private investment in nuclear power, a view that industry spokesmen and the Joint Committee shared. Because they recognized that the chances of a severe reactor accident could not be reduced to zero, even the most enthusiastic industry proponents of atomic power were reluctant to push ahead without adequate liability insurance. Private insurance companies would offer up to $60 million in coverage per reactor, an amount that far exceeded what was available to any other industry in the United States. However, in the event of a serious accident, that amount of coverage seemed insufficient to pay claims for deaths, injuries, and property damage in areas surrounding the malfunctioning plant.

Therefore, industry executives sought a Government program to provide additional insurance protection. Consolidated Edison, Inc., Board of Directors Chairman H.R. Searing declared that although his company would proceed with the construction of Indian Point plant located near New York City, it would not load fuel and begin operation unless the insurance question were resolved. General Electric's Francis K. McCune went even further by telling the Joint Committee in 1957 that if Congress did not enact indemnity legislation, his company would stop work on Commonwealth Edison Company's Dresden Nuclear Power Station, then under construction. He suggested that without a Government insurance plan, the market for civilian atomic energy would collapse and vendors would withdraw from the field.

Spurred by the industry's concerns, both the AEC and the Joint Committee considered methods that the Government could use to provide additional liability insurance for reactor owners. Their efforts culminated in legislation introduced by Senator Anderson and Congressman Melvin Price that proposed that the Government underwrite $500 million of insurance beyond the $60 million available from private companies. The AEC initially opposed setting a specific upper limit on the amount because no reliable method existed to estimate the possible damages from a reactor accident. However, Anderson rather arbitrarily decided on the $500 million figure because he wanted to avoid giving industry a "blank check." The bill stipulated that Congress could authorize additional payments if necessary and also required reactor owners to contribute funds to the insurance pool as their plants were licensed. With strong support from the AEC and the industry, Congress passed the Price-Anderson Nuclear Industries Indemnity Act (Price-Anderson Act) in August 1957. In final form, the measure also included Anderson's reforms to the AEC's licensing procedures. Although the agency disliked Anderson's amendments, it accepted them to avoid jeopardizing or retarding approval of the indemnity bill. In effect, the Price-Anderson Act was a regulatory measure because it provided insurance protection to victims of a nuclear accident, but it was largely promotional in motivation. Industry, the AEC, and the Joint Committee believed that it would remove a serious obstacle to private atomic development.

The Growth of Nuclear Power

The PRDC case and the Price-Anderson Act clearly illustrated the AEC's emphasis on developmental rather than regulatory efforts. The precedence that the AEC gave to promoting the growth of nuclear power resulted from a number of considerations. The Atomic Energy Act of 1954 made the encouragement of the widespread use of atomic energy for peaceful purposes a national goal, but private industry was often hesitant to assume the costs and risks of development. Therefore, the AEC sought to persuade or induce private interests to invest in nuclear power. This endeavor seemed particularly urgent because of the intense pressure the Joint Committee placed on the agency to speed progress and its persistent threat to require the AEC to construct prototype plants if private firms failed to act promptly. One important way in which the AEC pursued its objective of private development was to write regulations designed to protect public safety without being overly burdensome to industry.

Safety questions were largely a matter of judgment rather than something concrete or quantifiable, and AEC officials found it easier to assume that such issues had been or would be satisfactorily resolved than to assume that reactors would be built. For example, when the Commission issued a construction permit for the PRDC fast breeder reactor, its vision of an advanced technology plant that showed the effectiveness of its power demonstration reactor program outweighed the reservations of ACRS. Although the AEC was aware of the implications that safety questions posed for the development of the technology, it was confident that nuclear science, in due time, would provide the answers to outstanding issues. In short, the desire for tangible signs of progress was more compelling than first resolving more ethereal safety issues.

The AEC's emphasis on stimulating atomic development did not mean that it was inattentive to safety issues. The regulations that the staff drafted shortly after passage of the Atomic Energy Act of 1954 reflected careful consideration of the best scientific information and judgment available at the time. The AEC recognized and publicly acknowledged the possibility of accidents in such a new and rapidly changing technology; it never offered absolute assurances that accidents would not occur. Nevertheless, it believed that compliance with its regulations would minimize the chances of a serious accident. The agency did not view its developmental efforts as more important than regulatory policies, but it clearly viewed the encouragement of industrial growth as more immediate need.

By 1962, the AEC's efforts to stimulate private participation in nuclear power development had produced some encouraging results. In a report to President John F. Kennedy, the agency proudly pointed out that in the short time since atomic technology had been opened to private enterprise, six "sizeable" power reactors had begun operation, and two of those reactors had been built without Government subsidies. Despite industry's lingering concerns about the costs of nuclear power relative to fossil fuels, the AEC's promotional and regulatory programs had fostered the initial growth of commercial nuclear power. The agency predicted that by the year 2000 nuclear plants might provide up to 50 percent of the Nation's electrical generating capacity. Despite the AEC's claims, the future of the nuclear industry remained precarious. The 14 reactors in operation or under construction were still far from being commercially competitive or technologically proven, and interest in further development among utilities was uncertain. Both the AEC and Joint Committee were acutely aware of, and deeply disturbed about, those uncertainties.

Radiation Protection

To make matters worse from the perspective of nuclear proponents, there were signs of increasing public opposition to, or at least concern about, nuclear power hazards. In the early days of nuclear power development, public attitudes toward the technology were highly favorable, as the few opinion polls on the subject revealed. Press coverage of nuclear power was also overwhelmingly positive. For example, an article in National Geographic in 1958 concluded that "abundant energy released from the hearts of atoms promises a vastly different and better tomorrow for all mankind." However, in the late 1950s and early 1960s, the public became more alert to, and anxious about, the hazards of radiation, stemming largely from a major controversy over radioactive fallout from nuclear weapons testing. One result was that the public became increasingly troubled about the risks of exposure to radioactivity from any source, including nuclear power.

Before World War II, the dangers of radiation were primarily a matter of interest and concern to a relatively small group of scientists and physicians. Within a short period of time after the discovery of x rays and natural radioactivity in the 1890s, scientific investigators concluded that exposure to radiation could cause serious health problems, ranging from loss of hair and skin irritations to sterility and cancer. Ignorance of the hazards of x rays and radium and the use of them for frivolous purposes led to tragic consequences for people who received large doses of radiation from these sources. As experience with, and experimental data on, the effects of radiation gradually accumulated, professionals developed guidelines to protect x-ray technicians and other radiation workers from excessive exposure.

In 1934, a recently formed American committee representing professional societies and x-ray equipment manufacturers recommended for the first time a quantitative "tolerance dose" of radiation of 0.1 roentgen per day of whole-body exposure from external sources. The roentgen was a unit of measurement that indicated the effects of gamma rays or x rays on cells. Committee members believed that levels of radiation below the tolerance dose were generally safe and unlikely to cause injury "in the average individual." The following year, an international radiation protection committee composed of experts from five nations took similar action. Neither body regarded its recommended tolerance dose as definitive because empirical evidence remained fragmented and inconclusive. However, they were confident that available information made their proposals reasonable and provided an adequate margin of safety for the relatively small number of individuals exposed to radiation in their jobs.

The bombing of Hiroshima signaled the dawn of the atomic age and made radiation safety a vastly more complex task for two reasons. First, nuclear fission created many radioactive isotopes that did not previously exist in nature. Professionals in the field of radiation protection had to evaluate the hazards of these new little-known radioactive substances instead of considering only x rays and radium. Second, the problem of radiation safety extended to significantly larger segments of the population who could be exposed to radiation resulting from the development of new applications of atomic energy. Radiation protection broadened from a medical issue of limited proportions to a public health issue of, potentially at least, major dimensions.

As a result of these drastically altered circumstances, scientific authorities reassessed their recommendations on radiation protection. They modified their philosophy pertaining to radiological safety by abandoning the previous concept of a "tolerance dose," which assumed that exposure to radiation below the specified limits was generally harmless. Experiments in genetics indicated that reproductive cells were highly susceptible to damage from even small amounts of radiation. By the early 1940s, most scientists had rejected the idea that exposure to radiation below a certain threshold was inconsequential, at least with respect to genetic effects. In 1946, the National Committee on Radiation Protection (NCRP), a U.S committee of radiation experts, took action that reflected the consensus of opinion by replacing the terminology of "tolerance dose" with "maximum permissible dose," which it thought better conveyed the principle that no quantity of radiation was certifiably safe. It defined the "permissible dose" as that which "in the light of present knowledge, is not expected to cause appreciable bodily injury to a person at any time during his lifetime." While acknowledging the possibility that an individual could suffer harmful effects from radiation in amounts below the allowable limits, NCRP emphasized that the permissible dose was based on the belief that "the probability of the occurrence of such injuries must be so low that the risk should be readily acceptable to the average individual."

Because of the growth of atomic energy programs and the substantial increase in the number of individuals working with radiation sources, NCRP had decided by 1948 to reduce its recommended occupational exposure limits to 50 percent of the 1934 level. The International Commission on Radiological Protection (ICRP), NCRP's international counterpart, adopted the same maximum permissible dose after World War II. The new maximum permissible whole-body dose that NCRP and ICRP recommended was 0.3 roentgens per 6day work week, which was measured by exposure of the "most critical" tissue in the blood-forming organs, the gonads, and the lens of the eye. Higher limits applied for less sensitive areas of the body. In addition to the levels established for exposure to x rays or gamma rays, NCRP and ICRP also issued the maximum permissible concentrations in air and water for a list of radioactive isotopes that give off alpha or beta particles, known as "internal emitters." Alpha and beta particles cannot penetrate vital human tissue from outside the body, but they can pose a serious health hazard if they enter the body through the consumption of contaminated food or water or the inhalation of contaminated air.

The allowable limits established by both NCRP and ICRP applied only to radiation workers. However, because of the genetic effects of radiation and the possibility that other people could be exposed in an accident or an emergency, each group also issued guidelines for larger segments of the population. Because of the greater sensitivity of young persons to radiation, NCRP recommended that the occupational maximum permissible dose be reduced by a factor of 10 for anyone under the age of 18. ICRP went further by proposing a limit of one-tenth of the occupational level for the general population. Neither organization had any legal authority or official standing, but because their recommendations reflected the findings and opinions of leading experts in the field of radiation protection, they had significant influence on Government agencies concerned with radiological safety. The AEC used NCRP's occupational limits in its own installations and, after passage of the 1954 Atomic Energy Act, in its regulations for licensees. The agency's radiation protec-

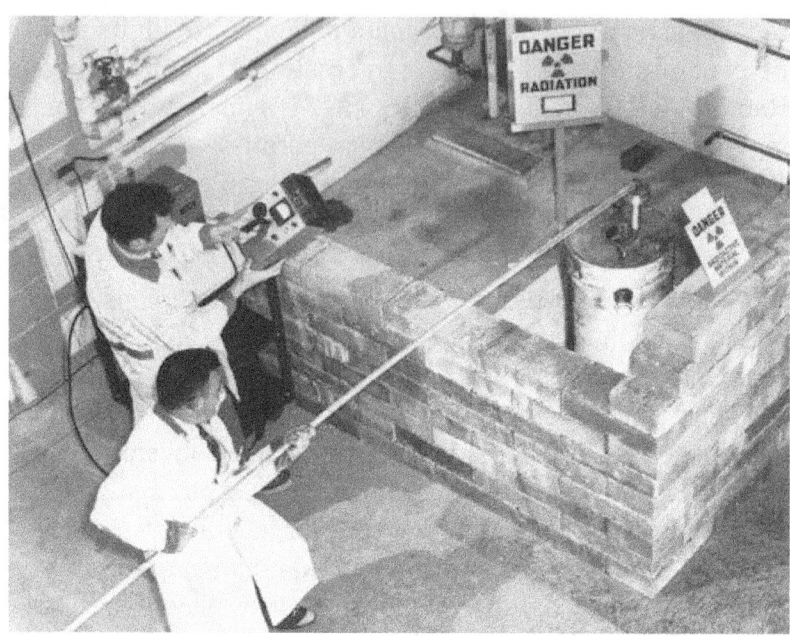

Researcher at Brookhaven National Laboratory removes plug from lead sheild containing radioactive materials Man at his left holds Geiger counter to monitor radiation levels

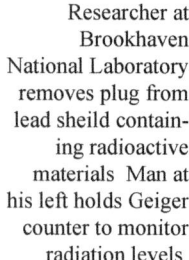

tion regulations, which were first issued for public comment in 1955 and became effective in 1957, followed NCRP's recommendations for radiation workers and set a permissible dose of one-tenth of the occupational level for members of the general population who potentially could be affected by licensee operations.

The Fallout Controversy

In the immediate postwar period, deliberations over the risks of radiation and permissible exposure levels were confined mostly to scientific circles. As a result of the fallout controversy, concern about radiation moved from the rarified realms of scientific and medical discourse to front page news. Atmospheric testing of nuclear weapons by the United States, the U.S.S.R., and Great Britain produced radioactive fallout that spread to populated areas far from the sites of the explosions. The fallout debate made radiation hazards a bitterly contested political issue. Scientists disagreed sharply about how serious a risk fallout presented to the general population, and this issue became a prominent subject in news reports, magazine stories, political campaigns, congressional hearings, and scientific studies. This issue not only focused public attention on the potential health hazards of relatively small amounts of radiation (as opposed to acute exposure) but also revealed that scientists did not know a great deal about the effects of low-level radiation.

The fallout controversy affected the AEC's regulatory program in two important ways. First, it led to a tightening of the agency's radiation standards. In response to increasing public concern and the findings of scientific groups, NCRP and ICRP both lowered their recommended permissible levels of exposure. Their actions provided a larger margin of safety, but they emphasized that no evidence existed to suggest that the previous levels had been dangerously high. They reduced their limits for occupational exposure to an average of 5 rem per year after the age of 18 while continuing to suggest that general population exposure levels be restricted to 10 percent of the occupational levels (0.5 rem per year) for individuals. The rem was a unit of measure that had largely replaced the roentgen and that indicated the biological effect of radiation exposure more precisely. For x rays and gamma rays, 1 rem equaled 1 roentgen. Radiation protection organizations added a new stipulation that, for genetic reasons, the average level for large population groups should not exceed one-thirtieth of the occupational limit, or 0.17 rem per year. The AEC promptly adopted the new recommendations as a part of its regulations; it issued them for public comment in 1959 and made them effective on January 1, 1961.

The fallout debate further influenced the AEC's regulatory program by arousing public anxieties about the health effects of low-level radiation. For example, the level of anxiety among members of the public was evident in citizen protests against the dumping of low-level radioactive wastes

in ocean waters. For more than a decade, the AEC had
authorized the dumping of such wastes under prescribed
conditions, but it became a subject of controversy only
after the fallout issue sensitized public opinion to radiation
hazards. In a similar manner, the first widespread objections
to the construction of proposed nuclear power plants arose
in the wake of the fallout debate. Citizen protests against
the construction of the Ravenswood plant in the heart of
New York City in 1963 and the Bodega Bay Nuclear Power
Plant on the coast of California near the boundary of the San
Andreas fault in 1963–1964 played a vital role in aborting
both projects.

At the end of the first decade following the passage of the
Atomic Energy Act of 1954, the prospects for rapid nuclear
power development were mixed. Impressive strides had cer-
tainly been made, but many uncertainties remained. Public
support for this technology was apparently strong, but this
support could not be taken for granted as the protests against
the Ravenswood and Bodega Bay plants had shown. However, beginning in the mid-1960s, a variety of considerations
fueled an unanticipated boom in the nuclear power industry
that resolved some of the unknowns about nuclear technology while at the same time raising a host of new questions
for the AEC's regulatory staff.

Chapter Two

The Nuclear
Power Debate,
1963–1975

During the late 1950s and early 1960s, the use of nuclear power to generate electricity was a novel and developing technology. Because relatively few plants were operating, under construction, or on order, the scope of the AEC's regulatory functions such as reactor siting, licensing, and inspection was still limited. However, during the later 1960s, the Nation's utilities rapidly increased their orders for nuclear power stations, participating in what Philip Sporn, past president of the American Electric Power Service Corporation, described in 1967 as the "great bandwagon market." At the same time, the size of nuclear plants that were under construction also expanded dramatically. The sudden arrival of commercially competitive nuclear power placed unprecedented demands on the AEC's regulatory staff and raised new safety problems that reactor experts had not previously considered. The surge in reactor orders and the growth in the size of individual plants also spurred new concerns about the environmental impact of nuclear power and intensified public uneasiness about the safety of the technology.

The Bandwagon Market

The bandwagon market was an outgrowth of several developments that enhanced the appeal of nuclear power to utilities in the mid- to late 1960s. One example was the intense competition between the two leading vendors of nuclear plants, General Electric and Westinghouse. In 1963, General Electric made a daring move to increase its reactor sales and to convince utilities that nuclear power was a safe, reliable, and cost-competitive alternative to fossil fuel. It offered a "turnkey" contract to Jersey Central Power and Light Company to build the 515megawatt electric (MWe) Oyster Creek Nuclear Generating Station near Toms River, NJ. For a fixed cost of $66 million, General Electric agreed to supply the entire plant to the utility. (The term "turnkey" suggested that the utility would merely have to turn a key to start operating the facility.) The company successfully outbid not only Westinghouse but also manufacturers of coal-fired units. General Electric expected to lose money on the Oyster Creek contract but hoped that the plant would help to stimulate the market for nuclear power.

The Oyster Creek contract opened the "turnkey era" of commercial nuclear power and came to symbolize the competitive debut of the technology. AEC Chairman Glenn T. Seaborg told President Lyndon B. Johnson that it represented an "economic breakthrough" for nuclear electricity. Westinghouse followed General Electric's lead in offering turnkey contracts for nuclear plants, setting off a fierce corporate battle. Turnkey plants were a financial blow for both companies; their losses ran into the hundreds of millions of dollars before they finally stopped offering turnkey arrangements. One General Electric official commented, "It's going to take a long time to restore to the treasury the demands we put on it to establish ourselves in the nuclear business." However, the turnkey contracts fulfilled General Electric's hopes of stirring interest among, and orders from, utilities. These contracts played a major role in triggering the bandwagon market.

Other important considerations at the time convinced a growing number of utilities to buy nuclear plants. One such consideration was the spread of power-pooling arrangements among utilities, which encouraged the construction of larger generating stations by easing fears of excess capacity and overexpansion. A utility with extra or reserve power could sell that power to other companies through interconnections. The desirability and feasibility of using larger individual plants worked to the benefit of nuclear vendors. They emphasized that bigger plants would produce economies of scale that would cut capital costs per unit of power and improve efficiency. This philosophy helped to overcome a major disadvantage of nuclear power relative to fossil fuel—the heavy capital requirements for building atomic plants. During the late 1960s, designs for nuclear facilities significantly increased from 500 MWe to 800 MWe to 1,000 MWe even though operating experience was still limited to units in the 200MWe range or less. The practice of "design by extrapolation" had been employed for fossil fuel units since the early 1950s. Before the mid-1960s, this approach appeared to work well, and vendors naturally extended it to nuclear units.

In addition to turnkey contracts, system interconnections, and an increase in unit size, a growing national concern about air pollution in the 1960s made nuclear power more attractive to utilities. Coal plants were major contributors to the deterioration of air quality and were obvious targets for cleanup efforts. As the campaign to improve the environment gained strength, the electric utility industry became more mindful of the cost of pollution control in fossil fuel plants. They increasingly viewed nuclear power as a good alternative to paying the expenses of pollution abatement in coal-fired units.

The bandwagon market for nuclear power reached its peak during 1966 and 1967, exceeding, in the words of one General Electric official, "even the most optimistic estimates." In 1965, the year before the reactor boom gathered momentum, nuclear vendors sold four nuclear plants with a total of 17 percent of the capacity that utilities purchased that year. In 1966, by contrast, utilities bought 20 nuclear units that made up 36 percent of the electrical capacity committed. The following year, nuclear vendors sold 31 units that represented 49 percent of the capacity ordered. In 1968, the number of reactor orders dropped to 17, but the percentage of the capacity filled with nuclear plants remained high at 47 percent.

The bandwagon market orders were large facilities that far exceeded the size of current operating reactors. Between 1963 (when the 515-MWe Oyster Creek reactor was ordered) and 1969 (when Oyster Creek began operation), the AEC issued 38 construction permits for units that were larger than Oyster Creek. Of those plants, 28 were in the 800- to 1,100-MWe range. The degree of extrapolation from small plants to mammoth ones was a matter of concern even to some strong nuclear advocates. By the late 1960s, it was apparent that design by extrapolation was not as successful as anticipated earlier for either nuclear or coal facilities. "We hoped the new machines would run just like the old ones we're familiar with," complained one utility executive about his huge coal-burning stations. However, he added that "they sure as hell don't."

Burdens of the Bandwagon Market

The rapid increase in the number of reactor applications and in the size of proposed plants placed enormous burdens on the AEC's regulatory staff. The flood of applications inevitably caused licensing delays because the AEC lacked enough qualified professionals. Between 1965 and 1970, the size of the regulatory staff increased by about 50 percent, but its licensing and inspection caseload increased by about 600 percent. The average time required to process a construction permit application stretched from about 1 year in 1965 to over 18 months by 1970. The growing backlog drew bitter complaints from utilities applying to build plants and from nuclear vendors. One utility executive predicted that if delays became commonplace, "it can safely be asserted that the splendid promise of nuclear power will have had a very short life." Another even more critical utility executive called the licensing process "a modern day Spanish Inquisition" carried out by "AEC engineers, scientists, and consultants [who] have no serious economic discipline." The AEC attempted to streamline its licensing procedures but found it impossible to reduce its review time or to satisfy the demands of the industry.

The licensing process became longer not only because of the number of applications that the AEC had to evaluate but also because of the complexity of the proposals that it received. The growth in the size of reactors and the practice of design by extrapolation raised many complex safety issues that could not be easily resolved. The exercise of careful judgment in assessing reactor applications was always critical, but it became even more so as utilities campaigned to build plants closer to populated regions. Although the AEC adopted an informal prohibition against "metropolitan siting" in urban locations (such as the proposed Ravenswood plant in downtown New York), it was more receptive to "suburban siting" fairly close to urban populations. This type of siting reduced the emphasis on one traditional means of protecting the public from the consequences of a nuclear accident— "remote siting." It placed greater dependence on another general method used to shield the public from the effects of an accident—engineered safeguards (a term that was later superseded by "engineered safety features") that were built into the plant. Even as the relative importance of engineered safeguards increased in the 1960s, questions arose about their reliability in preventing a massive release of radioactivity to the environment in the event of a severe accident.

The engineered safeguards in nuclear plants differed in design and operation, but they all performed two key functions: (1) to prevent the overheating and potential melting of the reactor core (which held the nuclear fuel) and (2) to prevent radioactive substances from escaping from the plant if the core was damaged. A number of systems were placed in reactors to remove heat and reduce excessive pressure if an accident occurred. For example, these systems included core sprays and pressure suppression pools; "safety injection" systems that would shoot large volumes of water into the reactor vessel; and combinations of filters, vents, scrubbers, and air circulators that would collect and retain radioactive gases and particles released during an accident. The final line of defense if the engineered safeguards failed was the containment building, a large, often dome-shaped structure that surrounded the reactor and the associated steam-producing equipment and safety systems.

Reactor experts were confident that the engineered safety features built into a plant and the containment structure would protect the public from the effects of an accident in almost any situation. However, they were troubled by the possibility that a chain of events could conceivably take place that would bypass or override the safety systems and, in the worst case, breach containment. "No one is in a position to demonstrate that a reactor accident with consequent escape of fission products to the environment will never happen," Clifford K. Beck, the AEC's Deputy Director of Regulation, told the Joint Committee in 1967. "No one really expects such an accident, but no one is in a position to say with full certainty that it will not occur."

The AEC strived to reduce the likelihood of an accident to a minimum. It based its decisions on the safety of reactor designs and plant applications on operating experience, engineering judgment, and experiments with test reactors. Experience with the first commercial reactors had been encouraging; it had provided a great deal of information that was useful in understanding reactor science. However, this experience was of limited application to the newer and larger reactors that utilities were building by the late 1960s. The rapid growth in reactor size placed a premium on the careful use of engineering judgment. To decrease the chances of a major accident that could threaten public health, the AEC required multiple backup equipment and redundancies in safety designs. It also employed conservative assumptions—that is, to assume the worst probable conditions for any postulated accident—about the ways in which an accident might damage or incapacitate safety systems in its evaluation of reactor proposals.

The Problem of Core Meltdown

The regulatory staff sought to gain as much experimental data as possible to enrich its knowledge and inform its collective engineering judgment. This was especially vital in light of the many unanswered questions about reactor behavior. The AEC had sponsored hundreds of small-scale experiments since the early 1950s that had yielded key information about a variety of reactor safety problems. However, these experiments provided little guidance on the issue of greatest concern to the AEC and the ACRS by the late 1960s—a core meltdown caused by a loss-of-coolant accident. Reactor experts had long recognized that a core meltdown was a plausible, if unlikely, occurrence. For example, a massive loss of coolant could occur if a large pipe that fed cooling water to the core broke. If the plant's emergency cooling systems also failed, the buildup of "decay heat," which resulted from continuing radioactive decay after the reactor shut down, could cause the core to melt. In older and smaller reactors, the experts were confident that even under the worst conditions—an accident in which the loss of coolant melted the core and it, in turn, melted through the pressure vessel that held the core—the containment structure would prevent a massive release of radioactivity to the environment. However, as proposed plants increased significantly in size, they began to worry that a core meltdown could lead to a breach of containment. This condition became their primary focus partly because of the greater decay heat that the larger plants would produce and partly because nuclear vendors did not increase the size of their containment buildings in corresponding proportions to the size of their reactors.

The greatest source of concern about a loss-of-coolant accident in large reactors was that the molten fuel would melt through not only the pressure vessel but also through the thick layer of concrete at the foundation of the containment

LOFT reactor under construction, 1969

building. The intensely radioactive fuel would then continue on its downward path into the ground. This scenario became known as the "China syndrome," because the melted core would presumably head through the Earth toward China. Other possible dangers of a core meltdown were that the molten fuel would breach containment either by reacting with water to cause a steam explosion or by releasing elements that could then combine to cause a chemical explosion. The precise effects of a large core meltdown were uncertain, but it was clear that the effects of radioactivity spewing into the atmosphere could be disastrous. ACRS and the regulatory staff regarded the chances of such an accident as low; they believed that it would occur only if the emergency core cooling system (ECCS), made up of redundant equipment that would rapidly feed water into the core, failed to function properly. However, they acknowledged the possibility that the ECCS might not work as designed. Without containment as a fail-safe final line of defense against any conceivable accident, they sought other means to provide safeguards against the China syndrome.

Chapter 2

At the prodding of ACRS, which first sounded the alarm about the China syndrome, the AEC established a special task force to look into the problem of core meltdown in 1966. The committee, chaired by William K. Ergen, a reactor safety expert and former ACRS member from Oak Ridge National Laboratory, submitted its findings to the AEC in October 1967. The report offered assurances about the improbability of a core meltdown and the reliability of ECCS designs, but it also acknowledged that a loss-of-coolant accident could cause a breach of containment if the ECCS failed to perform. Therefore, containment could no longer be regarded as an inviolable barrier to the escape of radioactivity. This finding represented a milestone in the evolution of reactor regulation. In effect, it imposed a modified approach to reactor safety.

Previously, the AEC had viewed the containment building as the final independent line of defense against the release of radiation; even if a serious accident took place, the damage that it caused would be restricted to the plant. However, once it became apparent that under some circumstances, the containment building might not hold, the key to protecting the public from a large release of radiation was to prevent accidents severe enough to threaten containment; the prevention of these types of accidents depended heavily on a properly designed and functioning ECCS.

The problem facing the AEC's regulatory staff was that experimental work and experience with emergency cooling was very limited. Finding a way to test and to provide empirical support for the reliability of emergency cooling became the central concern of the AEC's safety research program. Plans had been underway since the early 1960s to build an experimental reactor, known as the Loss-of-Fluid Tests (LOFT) reactor, at the AEC's reactor testing station in Idaho. Its purpose was to provide data about the effects of a loss-of-coolant accident. For a variety of reasons, including weak management of the test program, a change in design, and reduced funding, progress on the LOFT reactor and the preliminary tests that were essential for its success were chronically delayed. Despite the complaints of ACRS and the regulatory staff, the AEC diverted money from the LOFT project and other safety research projects on existing light-water reactor designs to other projects related to the development of fast breeder reactors. A proven fast breeder reactor was an urgent objective for the AEC and the Joint Committee; Seaborg described it as "a priority national goal" that could ensure "an essentially unlimited energy supply, free from problems of fuel resources and atmospheric contamination."

To the consternation of the AEC, experiments run at the Idaho test site in late 1970 and early 1971 suggested that the ECCS in light-water reactors might not work as designed. As a part of the preliminary experiments that were used to design the LOFT reactor, researchers ran a series of "semiscale" tests on a core that was only 9 inches long (compared to a length of 144 inches in a power reactor). The experiments were run by heating a simulated core electrically to allow the cooling water to escape and then injecting emergency coolant. To the surprise of the investigators, the high steam pressure that was created in the vessel by the loss of coolant blocked the flow of water from the ECCS. Without ever reaching the core, about 90 percent of the emergency coolant flowed out of the same break that had caused the loss of coolant in the first place.

In many ways, the semiscale experiments were not accurate simulations of designs or conditions in actual power reactors. The size, scale, and design of the experiments and the channels that directed the flow of coolant in the test model were markedly different from those in an actual reactor. Nevertheless, the results of the tests were disquieting. They introduced a new element of uncertainty into assessing the performance of ECCSs. The outcome of the tests had not been anticipated and called into question the analytical methods used to predict the events that would occur in a loss-of-coolant accident. These results were hardly conclusive, but their implications for the effectiveness of ECCSs were troubling.

The semiscale tests caught the AEC unprepared, and the AEC was uncertain about how to respond. Harold Price, the Director of Regulation, directed a special task force that he had recently formed to focus on the ECCS question and to draft a "white paper" within a month. Seaborg, for the first time, called the Office of Management and Budget to plead for more funds for safety research on light-water reactors. While waiting for the task force to finish its work, the AEC tried to keep information about the semiscale tests from getting into the public domain, even to the extent of withholding information about the tests from the Joint Committee. The results of the tests came at a very awkward time for the AEC. It was under renewed pressure from utilities facing power shortages and from the Joint Committee to streamline the licensing process and eliminate excessive delays. At the same time, Seaborg was successfully appealing to President Richard M. Nixon for support of the fast breeder reactor, and controversy over the semiscale tests and reactor safety could undermine congressional backing for the fast breeder reactor program. By spring 1971, nuclear critics were expressing opposition to the licensing of several proposed reactors, and news of the semiscale experiments seemed likely to support their efforts.

For those reasons, the AEC sought to resolve the ECCS issue as promptly and quietly as possible. It wanted to settle the uncertainties about safety without arousing a public debate that could slow the bandwagon market. Even before the task force that Price had established completed its study of the ECCS problem, the Commission decided to publish "interim acceptance criteria" for ECCSs that licensees would have to meet. It imposed a series of requirements that it believed would ensure that the ECCS in a plant would prevent a core meltdown after a loss-of-coolant accident. The AEC did not prescribe methods necessary for meeting the interim criteria, but, in effect, it mandated that manufacturers and utilities set an upper limit on the amount of heat generated by reactors. In some cases, this would force utilities to reduce the peak operating temperatures (and hence, the power) of their plants. Price told a press conference on June 19, 1971, that although the AEC thought that it was impossible "to guarantee absolute safety," he was "confident that these criteria will assure that the emergency core cooling systems will perform adequately to protect the temperature of the core from getting out of hand."

The interim ECCS criteria failed to achieve the AEC's objectives. News about the semiscale experiments triggered complaints about the AEC's handling of the issue even from friendly observers. It also prompted calls from nuclear critics for a licensing moratorium and a shutdown of the 11 plants then operating. Criticism expressed by the Union of Concerned Scientists (UCS), an organization that was established in 1969 to protest the misuse of technology and that had recently turned its attention to nuclear power, received wide publicity. UCS took a considerably less optimistic view of ECCS reliability than that of the AEC. It sharply questioned the adequacy of the interim criteria, charging, among other things, that they were "operationally vague and meaningless." Scientists at the AEC's national laboratories, without endorsing the alarmist language that UCS used, shared some of the same reservations. As a result of the uncertainties about ECCSs and the interim criteria, the AEC decided to hold public hearings that it hoped would help resolve the technical issues. It wanted to prevent the ECCS question from becoming a major impediment to the licensing of individual plants.

The AEC insisted that its critics had exaggerated the severity of the ECCS problem. The regulatory staff viewed the results of the failed semiscale tests as serious and took them into account when establishing the interim criteria. The regulatory staff also believed that the technical issues that the experiments raised would be resolved within a short period of time. It did not regard the tests as indications that existing designs were fundamentally flawed, and it emphasized the conservative engineering judgment it applied in evaluating plant applications. However, the ECCS controversy damaged the AEC's credibility and played into the hands of its critics. Instead of frankly acknowledging the potential significance of the ECCS problem and taking time to fully evaluate the technical uncertainties, the AEC acted hastily to prevent the issue from undermining public confidence in reactor safety or causing licensing delays. Its actions gave credence to the allegations of its critics that it was so determined to promote nuclear power and develop the fast breeder reactor that it was inattentive to safety concerns.

Thermal Pollution

By the time the ECCS issue hit the headlines, other questions about the environmental effects of nuclear power had eroded public support for the technology. The problem of industrial pollution and the deteriorating quality of the natural environment became increasingly urgent as a public policy issue during the 1960s. The increasing public and political concern about environmental protection, which occurred at a time when the demand for electricity was doubling every 10 years or more, placed utilities in a quandary. An article in Fortune magazine stated, "Americans do not seem willing to let the utilities continue devouring...ever increasing quantities of water, air, and land. And yet clearly they also are not willing to contemplate doing without all the electricity they want. These two wishes are incompatible. That is the dilemma faced by the utilities."

The Nuclear Power Debate, 1963–1975

Researchers from Argonne National Laboratory take measurements of the thermal discharge plume from the Big Rock Point Nuclear Power Plant on Lake Michigan

Utilities increasingly viewed nuclear power as the answer to that dilemma. Environmental concerns were a major incentive for the growth of the great bandwagon market, and nuclear power promised a means for meeting the demand for power without causing air pollution. Environmentalists recognized the benefits of nuclear power compared to fossil fuel, but they were more equivocal in their attitudes toward the technology than industry representatives were. Their ambivalence was perhaps best summarized by a statement from a leading environmental spokesman in 1967: "I think most conservationists may welcome the coming of nuclear plants, though we are sure they have their own parameters of difficulty."

Officials of the AEC actively promoted the idea that nuclear power provided the answer to both the environmental crisis and the energy crisis. Seaborg was especially outspoken on this point. Although he acknowledged that nuclear power could have some adverse impacts on the environment, he insisted that its effects were much less harmful than those of fossil fuel. In comparison with coal, he once declared,

Chapter 2

"There can be no doubt that nuclear power comes out looking like Mr. Clean."

In the late 1960s, a major controversy over the effects of waste heat from nuclear plants on water quality, widely known as "thermal pollution," undermined the view of nuclear power as beneficial to the environment relative to conventional fuels. Thermal pollution resulted from cooling the steam that drove the turbines to produce electricity in both fossil fuel and nuclear plants. The steam was condensed by the circulation of large amounts of water, and the cooling water was heated during the process, usually by 10 to 20 degrees Fahrenheit, before being returned to the body of water from which it came. This problem was not unique to nuclear plants, but it was more acute in them largely because fossil fuel plants used steam heat more efficiently than nuclear plants did. The problem of thermal pollution created more anxiety than during the 1960s because of the growing number of plants, the larger size of those plants, and the increasing inclination of utilities to order nuclear units.

Thermal pollution caused concern because it was potentially harmful to many species of fish. It could also disrupt the ecological balance in rivers and streams, allowing plants to thrive that made water look, taste, and smell unpleasant. Technical solutions to deal with thermal pollution were available, but they required extra costs in the construction and operation of steam-electric plants. For example, cooling towers of different designs or cooling ponds would greatly alleviate the release of waste heat to the source body of water. However, utilities resisted adding cooling apparatuses to the plants that they planned to build because of the expense and an appreciable loss of generating capacity.

Advocates in the news media for stronger Federal action to protect the environment, Congress, and State and Federal agencies urged the AEC to require its licensees to guard against the effects of thermal pollution. The AEC refused on the grounds that it lacked the statutory authority to impose regulations on hazards other than radiation. It argued that the Atomic Energy Act of 1954 restricted its regulatory jurisdiction to radiological dangers, a view that the U.S. Department of Justice and Federal courts upheld. This argument did not placate the AEC's critics, who accused it of ignoring a serious problem that nuclear plants exacerbated. Several members of Congress introduced legislation to grant the AEC authority over thermal pollution, but the agency opposed those measures unless fossil fuel plants were required to meet the same conditions. The AEC feared that nuclear power would be placed at a competitive disadvantage if plant owners had to provide cooling equipment that was not required for fossil-fuel-burning facilities.

The AEC came under increasing criticism for its position. The most prominent attack appeared in a Sports Illustrated article in January 1969. It assailed the AEC for failing to regulate against thermal pollution and attributed its inaction to a fear of the "financial investment that power companies would have to make…to stop nuclear plants from frying fish or cooking waterways wholesale." The article was a distorted and exaggerated presentation, but it contributed to a growing perception that instead of being a solution to the dilemma of producing electricity without causing serious environmental damage, nuclear power was a part of the problem.

Eventually, the controversy over thermal pollution died out. One reason was that Congress passed legislation that gave the AEC authority to regulate against thermal pollution and that applied to most fossil fuel plants as well. A more important reason was that utilities increasingly took action to curb the consequences of discharging waste heat. Although they initially resisted the calls for cooling equipment, they soon found that the costs of responding to litigation, enduring postponements in the construction or operation of new plants, or suffering a loss of public esteem were less tolerable than those of building cooling towers or ponds. By 1971, most nuclear plants being built on, or planned for, inland waterways (where the problem was most acute) included cooling systems. However, the legacy of the thermal pollution debate lingered on. This concern undermined confidence in the AEC and awakened public doubts about the environmental impact of nuclear power. The thermal pollution debate played a vital role in transforming the initial ambivalence that environmentalists had demonstrated toward the technology into strong and vocal opposition. As a result of the thermal pollution issue, the AEC and the nuclear industry frequently found themselves included among the ranks of enemies of the environment.

The Radiation Debate

The thermal pollution question was the first, but not the only, debate over the effects of nuclear power that aroused widespread public concern in the late 1960s and early 1970s. A major controversy that arose over the effects of low-level radiation from the routine operation of nuclear plants also fed fears about the expanding use of the technology. Drawing on the recommendations of NCRP, the AEC had established limits for public exposure to radiation from nuclear plants of 0.5 rem per year for individuals. To determine the allowable release of radioactive effluents from a plant, it assumed that a person stood outdoors at the boundary of the facility 24 hours a day, 365 days a year. Licensees generally met the requirements easily. In 1968, for example, releases from most plants measured less than 3 percent of the permissible levels for liquid effluents and less than 1 percent for gaseous effluents. The conservative assumptions of the AEC and the performance of operating plants did not prevent criticism of the AEC's radiation standards. A number of observers suggested that the AEC's regulations were insufficiently rigorous and should be substantially revised because of the uncertainties about the effects of low-level radiation. This suggestion first emerged as a widely publicized issue when the State of Minnesota, responding to questions raised by environmentalists, stipulated in May 1969 that a plant under construction must restrict its radioactive effluents to a level of about 3 percent of that allowed by the AEC.

The adequacy of the AEC's radiation standards became even more contentious in the fall of 1969, when two prominent scientists, John W. Gofman and Arthur R. Tamplin, suggested that if everyone in the United States received the permissible population dose of radiation, it would cause 17,000 (later revised to 32,000) additional cases of cancer annually. Gofman and Tamplin worked at Lawrence Livermore National Laboratory, which was funded by the AEC; their position as insiders therefore gave their claims special credibility. They initially proposed that the AEC lower its limits by a factor of 10 and later urged that it require a zero release of radioactivity.

Gofman and Tamplin not only argued that the existing standards of the AEC and other radiation protection organizations were inadequate but also challenged the prevailing consensus that the benefits of nuclear power were worth the risks. Gofman was especially harsh in his analysis; he insisted that "the AEC is stating [in its radiation protection regulations] that there is a risk and their hope that the benefits outweigh the number of deaths." He added, "This is legalized murder, the only question is how many murders."

The AEC denied Gofman's and Tamplin's assertions on the grounds that they had extrapolated from high doses to estimate the hazards of low-level exposure and that it was impossible for the entire Nation to receive the levels of radiation that applied at plant boundaries. Most authorities in the field of radiation protection agreed with the AEC that the risks of effluents from nuclear power were far smaller than those maintained by Gofman and Tamplin. Nevertheless, in an effort to provide an extra measure of protection, reassure the public, and undercut the appeal of its critics, the AEC issued for public comment in June 1971 new "design objectives" for nuclear plants that would, in effect, reduce the permissible levels of effluents by a factor of about 100. This action elicited protests from industry representatives and from radiation protection professionals, but it did not impress many critics who expressed doubt that the AEC would enforce the new guidelines. The controversy focused public attention, once again, on the effects of low-level radiation, but it did little to clarify a complex and ambiguous issue.

The National Environmental Policy Act and Calvert Cliffs Nuclear Power Plant

In addition to the objections that its positions on thermal pollution and radiation standards stirred, the AEC provoked sharp criticism for its response to the National Environmental Policy Act (NEPA). The law, passed by Congress in December 1969 and signed by President Nixon on January 1, 1970, required Federal agencies to consider the environmental impact of their activities. The measure was in many ways vague and confusing, and it gave Federal agencies broad discretion in deciding how to carry out this mandate. The AEC acted promptly to comply with NEPA, but its procedures for doing so brought protests from environmentalists. The agency took a narrow view of its responsibilities under NEPA. In a proposed regulation that the agency issued in December 1970, it included, for the first time, nonradiological issues in its regulatory jurisdiction. However, the AEC also stipulated that it intended to rely on the environmental assessments of other Federal and State agencies (rather than conducting its own), it agreed to consider environmental issues in licensing board hearings only if a party to the proceeding raised these issues, and it postponed any review of NEPA issues in licensing cases until March 1971.

The AEC declined to take an expansive view of its responsibilities under NEPA for several reasons. One was the conviction that the routine operation of nuclear plants was not a serious threat to the environment and, indeed, was beneficial compared to burning fossil fuels. Other legislation covered the major products of nuclear power generation that affected the environment, radiation releases and thermal discharges. Furthermore, implementation of NEPA might divert the AEC's limited human resources from tasks that were more central to its mission. The regulatory staff was "all but overwhelmed" by the flood of reactor applications and did not relish the idea of having to spend large amounts of time on environmental reviews. Most importantly, the AEC feared that weighing environmental issues other than radiation and thermal releases would cause unwarranted delays in licensing plants. The time required for evaluating applications was already increasing, and the AEC worried that NEPA could force a "quantum leap" in the length of the process. It sought to strike a balance between environmental concerns and the need for electrical power in framing its regulations.

Environmentalists complained that the AEC had failed to fulfill the purposes of NEPA and took the agency to Federal court over the application of the AEC's regulations to the Calvert Cliffs Nuclear Power Plant, then under construction on the Chesapeake Bay in rural Maryland. On July 23, 1971, the U.S. Court of Appeals for the District of Columbia handed down a ruling that was a crushing defeat for the AEC. The court sternly rebuked the agency in its most widely quoted statement: "We believe that the Commission's crabbed interpretation of NEPA makes a mockery of the Act." The Calvert Cliffs decision was, in the words of

Nucleonics Week, a "stunning body blow" to the AEC and the nuclear industry.

The Calvert Cliffs decision was another in a series of setbacks for the AEC and nuclear power. It was apparent by the summer of 1971 that public distrust of the AEC was growing and support for nuclear power was declining. The cumulative effect of controversies over ECCS, thermal pollution, radiation standards, NEPA, and other issues eroded public confidence in the AEC's commitment to safety and raised doubts about the benefits of nuclear power. Antinuclear activists capitalized on growing uneasiness about the health and environmental effects of the technology. Some of the critics were well informed and responsible in their arguments, but others were one sided and inaccurate. Attempts by nuclear proponents to correct a plethora of misleading and exaggerated stories, advertisements, speeches, and other presentations inevitably failed to win as much attention or produce the same effect. To make matters worse for the AEC, it suffered from the general disillusionment with the Government, established institutions, and science that prevailed by the late 1960s, largely as a result of the Vietnam war. One college student summarized the situation after listening to a debate between Victor Bond, a radiation expert from Brookhaven National Laboratory, and a vocal AEC critic: "Dr. Bond sounds good, but we can't believe him. He works for the government."

By the summer of 1971, the AEC was an embattled agency, largely, though not exclusively, because of regulatory issues. Seaborg, after serving as chairman for 10 years, resigned his post in July 1971, and President Nixon appointed James R. Schlesinger, Assistant Director of the Office of Management and Budget, to take his place. Schlesinger was determined to make the AEC more responsive to environmental concerns and to improve its tarnished public image. As an important first step in those efforts, he and William O. Doub, who took a seat on the Commission at the same time that Schlesinger assumed the chairmanship, concluded that the AEC should not appeal the Calvert Cliffs ruling, and, after considering the alternatives, their colleagues agreed. The AEC announced its decision on August 26, 1971.

The AEC's response to the Calvert Cliffs decision brought a storm of protests from utilities that feared long delays in the licensing of plants that were nearly ready for operation. Schlesinger explained the AEC's new position in a speech he delivered to a meeting of industrial groups in Bal Harbour, FL, on October 20, 1971. He told his audience that although the long-term outlook for nuclear power appeared "bullish," the pace of development depended on two variables: "first, the provision of a safe, reliable product; second, achievement of public confidence in that product." Schlesinger declared that the AEC's policy of promoting and protecting the industry had been justified to help nuclear power get started, but because the industry was "rapidly approaching mature growth," the AEC must redefine its responsibilities. "You should not expect the AEC," he announced, "to fight the industry's political, social, and commercial battles." He added that the agency's role was "primarily to perform as a referee serving the public interest." The message that Schlesinger's speech conveyed was unprecedented; it proclaimed a sharp break with the AEC's history and a new direction in the agency's approach to its regulatory duties.

Schlesinger's efforts to narrow the divisions between nuclear proponents and critics and to recover the AEC's regulatory credibility produced, at best, mixed results. Many environmentalists were pleased with the AEC's acceptance of the Calvert Cliffs ruling and with Schlesinger's Bal Harbour speech. Their guarded optimism about Schlesinger's attitudes was perhaps best summarized by the title of an article about him in National Wildlife magazine: "There's a Bird Watcher Running the Atomic Energy Commission." However, major differences between the AEC and environmentalists remained; many of the same issues that had aroused concern before Schlesinger's arrival continued to generate controversy.

New Controversies and the End of the Atomic Energy Commission

The reliability of ECCSs was an issue that continued to generate controversy. In light of the objections to the interim acceptance criteria for ECCSs that the AEC had published in June 1971, the agency decided to hold a rulemaking hearing on the issue that would apply to all licensing cases. It hoped that this would avoid repeating the same procedures and deliberating over the same questions in case-by-case hearings and that generic hearings would provide a means to resolve issues common to all plants. The ECCS hearings got underway in early 1972 and amounted to 135 days over a period of one-and-a-half years. When the hearings ended, the transcripts of the proceedings filled more than 22,000 pages. The ECCS hearings led to a final rule that made some small but important revisions in the interim criteria. They also produced acrimonious testimony and front-page headlines that often reflected unfavorably on the AEC's safety programs, revealed divisions among its own experts about the value of the interim criteria, and further damaged its credibility.

Another issue that undermined confidence in the AEC in the early 1970s was its approach to high-level radioactive waste disposal. The growth of the nuclear power industry made the safe disposal of intensely radioactive waste materials an increasingly urgent matter. The AEC had investigated the means of dealing with reactor wastes for years, but it had not found a solution to the problem. As early as 1957, a scientific consensus had concluded that deep underground salt deposits were the best repositories for long-lived and highly radioactive wastes. In 1970, in response to increasing expressions of concern about the lack of a policy for high-level waste disposal from scientific authorities, members of Congress, and the press, the AEC announced plans to develop a permanent repository for nuclear waste in an abandoned salt mine near Lyons, KS. It aired its plans without conducting thorough geologic and hydrologic investigations, and the suitability of the site was soon challenged by the State geologist of Kansas and other scientists. The uncertainties about the site generated a bitter dispute between the AEC on one side and members of Congress and State officials from Kansas on the other. The dispute ended in 1972 in great embarrassment for the AEC when the concerns of those who opposed the Lyons, KS, location proved to be well founded.

In addition to debates over the potential failure of ECCSs and high-level waste disposal, questions concerning reactor design and safety, quality assurance (QA), the probability of a major reactor accident, and other issues fueled the controversy over nuclear power. The number of contested hearings for plant licenses steadily grew. The ongoing controversy frustrated Schlesinger's hopes of increasing public confidence in the AEC and of defusing the conflicts between opposing views. By highlighting the issues on which the AEC's performance was suspect, the agency also obscured the requirements that its regulatory staff imposed over the protests and against the wishes of the nuclear industry, the high standards that it demanded in the design and construction of nuclear plants, and the conservative assumptions that it applied in evaluating plant applications and formulating radiation protection regulations.

As the nuclear power debate continued, the AEC came under increasing attacks for its dual responsibilities for developing and regulating the technology. This issue became a major argument that nuclear critics cited in their indictments of the AEC. One critic said that it was "like letting

the fox guard the henhouse." The possibility of creating
separate agencies to promote and to regulate the civilian
uses of nuclear energy had arisen within a short time after
passage of the Atomic Energy Act of 1954. However, in the
early stages of nuclear development, this possibility had
seemed premature and unwarranted. The idea of creating
separate agencies gained greater support as both industry
concerns and antinuclear sentiment grew, and it took on
greater urgency after the Arab oil embargo and the energy
crisis of 1973–1974. One of President Nixon's responses to
the energy crisis was to ask Congress to create a new agency
that could focus on, and presumably speed up, the licensing
of nuclear plants. After much debate, Congress divided the
AEC into the U.S. Energy Research and Development Administration and the U.S. Nuclear Regulatory Commission
(NRC) in legislation that it passed in 1974. The Energy Re-
organization Act of 1974, coupled with the Atomic Energy
Act of 1954, constituted the statutory basis for the NRC.
The new agency inherited a mixed legacy from its predeces-
sor, marked both by 20 years of conscientious regulation
and by unresolved safety questions, substantial antinuclear
activism, and growing public doubts about nuclear power.

Chapter Three

The U.S. Nuclear Regulatory Commission and Three Mile Island

Chapter 3

The NRC began its operations as a separate agency in January 1975. In many ways, it carried on the legacy inherited from the AEC. It performed the same licensing and rule-making functions that the regulatory staff had discharged for two decades. It also assumed some new administrative and regulatory duties. The NRC, unlike the AEC's regulatory staff, was the final arbiter of regulatory issues; its judgment on safety questions was less susceptible to being overridden by developmental priorities. This did not mean that the NRC acted without regard to industry concerns or that its officials always agreed on policy matters, but it did mean that the agency's statutory mandate was clearly focused on ensuring the safety of nuclear power.

The NRC devoted a great deal of attention during its first few months to organizational tasks at the same time that it carried out its regulatory responsibilities. It deliberated over a number of pressing problems that it inherited from the AEC or that arose shortly after its establishment. One issue that received particular attention was the safeguarding of nuclear materials. The term "safeguards" applied to the prevention of theft, loss, or diversion of nuclear fuel or other materials or the sabotage of nuclear plants. This question took on greatly increased importance and visibility in the early 1970s because of growing apprehension about the activities and intentions of terrorist groups. There was a wave of terrorist bombings, assassinations, hijackings, and murders at that time, perhaps the most shocking of which was the murder of Israeli athletes at the 1972 Olympics.

The increase in such attacks around the world raised new concerns that terrorists would be able to build an atomic bomb, which was underscored by the well-publicized warnings of some nuclear experts that making a bomb was not terribly difficult for anyone who obtained the necessary materials. As a result, the AEC, and, after its abolition, the NRC, substantially strengthened regulatory requirements for

The U.S. Nuclear Regulatory Commission and Three Mile Island

the transportation of nuclear materials and for nuclear plant security. The NRC also devoted considerable attention to the export of nuclear materials to foreign countries. The United States was by far the leading supplier of nuclear fuel and other materials for the production of nuclear power abroad, and the NRC exercised important responsibilities for ensuring that nuclear exports did not encourage the proliferation of nuclear weapons or make them available to terrorists.

Despite the prominence of safeguards problems, the central issue for the NRC at the time of its creation remained reactor safety. Two events occurred in the early months of the NRC's existence that commanded the particular attention of the agency and the public. The first event was a major fire at the Tennessee Valley Authority's Browns Ferry Nuclear Plant near Decatur, AL, in March 1975. In the process of looking for air leaks in an area containing trays of electrical cables that operated the plant's control room and safety systems, a technician set off a fire. He used a lighted candle to conduct the search, and the open flame ignited the insulation around the cables. The fire raged for over 7 hours and nearly disabled the safety equipment of one of the two affected units. The accident was a blow to the public image of nuclear power and the recently established NRC. It focused new attention on preventing fires that threaten plant safety and on the possibility of "common-mode failures" in which a single breakdown could initiate a chain of events that incapacitated even the redundant safety features.

The second source of unusually extensive discussion and considerable controversy shortly after the NRC began operations was the publication of the final version of the "Reactor Safety Study" that the AEC had commissioned in 1972. The purpose of the study was to estimate the probability of a severe reactor accident—an issue that the AEC had never found a satisfactory means of addressing. To direct the study, the AEC had recruited Norman C. Rasmussen, a professor of nuclear engineering at the Massachusetts Institute of Technology (MIT). Rasmussen, assisted by AEC staff members, applied new methodologies and sophisticated "fault-tree analyses" to project the likelihood of a serious nuclear accident. The final Rasmussen report, released in October 1975, concluded that risks from nuclear power were very small in comparison to other risks from, for example, fires, explosions, toxic chemical spills, dam failures, airplane crashes, earthquakes, tornadoes, and hurricanes.

Although the Rasmussen report was hailed as a pioneering effort that enlightened a complex subject, it also drew criticism from both inside and outside the NRC. Some authorities suggested that the study failed to account for the many paths that could lead to major accidents. Others complained that the data in the report did not support its executive summary's conclusions about the relative risks of nuclear power. After considering the arguments on both sides of the issue, the Commission issued a policy statement in January 1979 that withdrew its full endorsement of the study's executive summary.

The Three Mile Island Accident

Within a short time, the discussion of severe nuclear accidents ceased to be strictly a matter of theoretical projections. On March 28, 1979, an accident at the Three Mile Island Nuclear Station (TMI), Unit 2, near Harrisburg, PA, made the issue starkly and alarmingly real. As a result of a series of mechanical failures and human errors, the accident (researchers later determined) uncovered the reactor's core and melted about half of it. The immediate cause of the accident was a pressure relief valve that stuck open and allowed large volumes of reactor coolant to escape from the core. The control room instrument panel did not provide a clear picture of what was happening in the reactor, and, partly as a result, the plant's operators failed to pick up the signs of a loss-of-coolant accident. Although the ECCSs began

The U.S. Nuclear Regulatory Commission and Three Mile Island

Three Mile Island, looking southeast The accident occurred in the reactor at the right of the photograph

to work according to design, the operating crew decided to reduce the flow from them to a trickle. By the time that experts realized that the plant had undergone a loss-of-coolant accident and flooded the core, the reactor had suffered irreparable damage.

The credibility of the nuclear industry and the NRC fared almost as badly. Uncertainty about the causes of the problem, confusion about how to deal with it, conflicting information from Government and industry experts, and contradictory appraisals about the level of danger in the days following the accident fed public fears and fostered a deepening perception of a technology that was out of control. The greatest source of concern was a hydrogen bubble that formed in the pressure vessel, the large container that held the reactor core. At first, experts feared that the bubble could inhibit efforts to cool the core and bring it to a safe-shutdown condition. However, another issue soon arose. Joseph M. Hendrie, Chairman of the NRC, began to worry that over time the bubble might become flammable or even explosive. In a worst case scenario, a burn or explosion could rupture the pressure vessel and increase by indeterminate but uncomfortable proportions the chances of a breach of containment and a massive release of radiation to the environment.

Hendrie immediately instructed the NRC staff to explore the possibility that the bubble could reach a flammable or explosive condition. News of the bubble created a great deal of apprehension among the population who lived near the plant, and thousands of people evacuated their homes as headlines warned that a "hydrogen explosion" could occur. They joined those who had left the area the previous day in response to a voluntary advisory evacuation for pregnant women and preschool-aged children that Governor Richard Thornburgh had issued. He had acted in consultation with the NRC in response to the existing uncertainties about the level of danger from the accident. Over a 5-day period, about 144,000 people evacuated the area with remarkable calmness and responsibility.

While the NRC investigated the bubble problem, Thornburgh called a late-night press conference. Harold R. Denton, the NRC's chief staff official at the site, explained that the bubble could conceivably be a hazard in a matter of days but that it did not pose an immediate threat. Denton's assurances curbed the sense of alarm among the local population that the plant might suddenly explode. The following day, after a highly publicized tour of the plant by President Jimmy Carter, the NRC determined that a lack of oxygen in the pressure vessel prevented the bubble from reaching a flammable or explosive state. This conclusion, along with the gradual reduction in the size of the bubble, ended the acute phase of the TMI crisis. However, many serious questions about the safety of nuclear power remained to be addressed.

In some ways, the TMI accident produced reassuring, or at least encouraging, information for reactor experts about the design and operation of the safety systems in a large nuclear plant. Despite the substantial degree of core melting that occurred, containment was not breached. From all indications, the amount of radioactivity released into the environment as a result of the accident was very low. For example, less than 20 curies of the 66 million curies of iodine131 in the reactor at the time of the accident escaped from the plant. Careful epidemiological studies of the population in the region surrounding the plant revealed no increase in the incidence of cancer over a period of two decades that could be attributed to the accident.

The favorable findings about the effects of the accident were overshadowed by a series of unsettling disclosures of problems that demanded immediate correction. The TMI accident focused attention on possible causes of accidents that the AEC/NRC and the nuclear industry had not considered extensively. Their working assumption had been that the most likely cause of a loss-of-coolant accident was a break in a large pipe that fed coolant to the core. However, the destruction of the core at TMI had not resulted from a large pipe break but instead from a relatively minor mechanical failure that operator errors had drastically compounded.

Perhaps the most distressing revelation of TMI was that an accident so severe could occur at all. Neither the AEC/NRC nor the industry had ever claimed that a major reactor accident was impossible despite the multiple and redundant safety features that are built into nuclear plants. However, they had regarded it as highly unlikely, to the point of being nearly incredible. The TMI accident demonstrated graphically that serious consequences could arise from unanticipated events. This enhanced the credibility of nuclear critics who had argued for years that no facility as complex as a nuclear plant could be made foolproof. Public opinion polls taken after the TMI accident showed a significant erosion in support for nuclear power. One survey found that for the first time, the number of respondents who opposed building more nuclear units exceeded those who favored new plants. However, the polls indicated that the public did not want to abandon nuclear power or close existing plants.

The NRC Response to the Accident at Three Mile Island

The NRC responded to TMI by reexamining the adequacy of its safety requirements and by imposing new regulations to correct deficiencies. It placed much greater emphasis on "human factors" in plant performance in an effort to avoid a repeat of the operator errors that had exacerbated the accident. The agency developed more stringent requirements for operator training, testing, and licensing. In cooperation with industry groups, it promoted the increased use of reactor simulators and the careful assessment of control rooms and instrumentation. In addition, the agency expanded its resident inspector program to station at least two of its inspectors at each plant site.

The NRC devoted greater attention to other problems that had received limited consideration before the TMI accident. These problems included the possible effects of small failures that could lead to major consequences, such as those that happened at TMI. The agency sponsored a series of studies on the ways in which "small breaks and transients" could threaten plant safety. A second area of NRC focus was the evaluation of operational data from licensees. It established a new office to systematically review information from, and the performance of, operating plants. This action reflected the belated recognition that malfunctions similar to those at TMI had occurred at other plants, but the information had never been assimilated or disseminated.

The NRC undertook other initiatives as a result of TMI. It decided to review radiation protection procedures at operating plants to assess their adequacy and to look for ways to improve existing regulations. It expanded research programs on problems that TMI had highlighted, including fuel damage, fission-product release, and hydrogen generation and control. In light of the confusion and uncertainty over the evacuation of the areas surrounding TMI during the accident, the NRC also sought to upgrade emergency preparedness and planning. Those and other steps that it undertook in the wake of the accident were intended to reduce the likelihood of a major accident and, in the event that one occurred, to enhance the ability of the NRC, the utility, and the public to cope with it.

Chernobyl

While the NRC was still deliberating over and revising its requirements in the aftermath of TMI, another event shook the industry and further undercut public support for nuclear power. This time, the NRC was a distant though interested observer rather than a direct participant. On April 26, 1986, Unit 4 of the nuclear power station at Chernobyl in the U.S.S.R. underwent a violent explosion that destroyed the reactor and blew the top off it, spewing massive amounts of radioactivity into the environment. The accident occurred during a test in which operators had turned off the plant's safety systems and then lost control of the reactivity in the reactor. Without emergency cooling or a containment building to stop or at least slow the escape of radiation, the areas around the plant quickly became seriously contaminated, and a radioactive plume spread far into other parts of the U.S.S.R. and Europe. Although the radiation did not pose a threat to the United States, one measure of its intensity in the U.S.S.R. was that the levels of iodine-131 around TMI were three times as high after the Chernobyl accident than they had been after the TMI accident.

The design of the Chernobyl reactor was entirely different than that of U.S. plants, and the series of operator blunders that led to the accident defied belief. Supporters of nuclear power emphasized that a Chernobyl-type accident could not occur in commercial plants in the United States (or other nations that used U.S. designs) and that U.S. reactors featured safety systems and containment to prevent the release of radioactivity. However, nuclear critics pointed to Chernobyl as the prime example of the hazards of nuclear power. A representative of UCS remarked, "The accident at Chernobyl makes it clear. Nuclear power is inherently dangerous." A popular slogan that quickly appeared on the placards of European environmentalists was "Chernobyl Is Everywhere." The Chernobyl tragedy was a major setback for nuclear proponents in their hope to win public support for the technology and to spur orders for new reactors. For example, a poll conducted in May 1986 found that 78 percent of respondents opposed the construction of more nuclear plants in the United States. Utilities had not ordered any new plants since 1978, and the number of cancellations for planned units was growing. "We're in trouble," conceded a spokesman for the Atomic Industrial Forum, Inc. "If the calls I have received from people in the industry are a good indication, they are all very worried."

Licensing of New Plants and Emergency Planning

The Chernobyl accident added a new source of concern to longstanding controversies over the licensing of several reactors in the United States. In the aftermath of the TMI accident, the NRC had suspended the granting of operating licenses for plants that were in the pipeline.

This "licensing pause" for fuel loading and low-power testing ended in February 1980. In August 1980, the NRC issued the first full-power operating license (to North Anna Power Station, Unit 2, in Virginia) since the TMI accident. In the following 9 years, it granted full-power licenses to over 40 other reactors, most of which had received construction per-

The U.S. Nuclear Regulatory Commission and Three Mile Island

mits in the mid1970s. In 1985, it authorized the undamaged TMI, Unit 1, which had been shut down for refueling at the time of the accident at TMI, Unit 2, to resume operation.

Although many of the licensing actions aroused little opposition, others triggered major controversies. The two licensing cases that precipitated what were perhaps the most bitter, protracted, and widely publicized debates were Seabrook Nuclear Power Plant in New Hampshire and Shoreham Nuclear Power Plant on Long Island, NY. The key, although hardly the sole, issue in both cases was emergency planning. The TMI accident had vividly demonstrated the deficiencies in existing procedures for coping with an offsite nuclear emergency. The lack of effective preparation had produced confusion and uncertainty among decisionmakers and among members of the public faced with the prospect of exposure to radiation releases from the plant. After the accident, the NRC, prodded by Congress to improve emergency planning, adopted a new rule on emergency planning. It required each nuclear utility to come up with a plan for evacuating the population within a 10mile radius of its plant(s) in the event of a reactor accident, although protective action was likely to be necessary only in a part of the "emergency planning zone." The rule applied to plants in operation and under construction. It called for plant owners to work with State and local police, fire, and civil defense authorities to put together an emergency plan that the NRC and the Federal Emergency Management Agency would test and evaluate.

 The NRC lacked the authority to force State and local governments to participate in emergency preparedness procedures, and it had little choice but to frame its regulations on the assumption that they would cooperate. The agency recognized that State or local governments, if they chose to, could try to prevent the operation of nuclear plants by refusing to work with Federal agencies to improve emergency planning. That was precisely what the States of New York

Chapter 3

Opponents of a full-power license for Seabrook express their views at NRC headquarters in Rockville, Maryland, 1990

and Massachusetts sought to do in the cases of Shoreham and Seabrook. In New York, Governor Mario M. Cuomo and other state officials claimed that it would be impossible to evacuate Long Island if Shoreham suffered a major accident. Although plant proponents pointed out that emergency plans did not require the evacuation of all of Long Island if a serious accident occurred, the State refused to join in emergency planning procedures or drills. The NRC granted Shoreham a low-power operating license, but the State and the utility, Long Island Lighting Company, eventually reached a settlement in which the company agreed not to operate the plant in return for concessions from the State.

A similar issue arose at Seabrook, although the outcome was different. The plant is located in the State of New Hampshire, but the 10mile emergency planning zone extended across the State line into Massachusetts. By the time that construction of the plant was completed, Massachusetts Governor Michael S. Dukakis, largely as a result of Chernobyl, had decided that he would not cooperate with emergency planning efforts for Seabrook. New Hampshire officials worked with Federal agencies to prepare an emergency plan, but Massachusetts refused to cooperate, arguing that crowded beaches near the Seabrook plant could not be evacuated in the event of an accident. As a result of New York's and Massachusetts' positions on Shoreham and Seabrook, the NRC adopted a "realism rule" in 1988 that was grounded on the premise that in an actual emergency, State and local governments would make every effort to protect public health and safety. Therefore, in cases in which State or local officials declined to participate in emergency planning, the NRC and the Federal Emergency Management Agency would review and evaluate plans developed by the utility. On that basis, the NRC issued an operating license for the Seabrook plant. The arguments that raged over emergency planning and other issues at Shoreham and Seabrook attracted a great deal of attention, spawned heated controversy, and raised anew an old question about the relative authority of Federal, State, and local governments in licensing and regulating nuclear plants.

The lengthy and laborious licensing procedures that applicants had to undergo in the cases of Shoreham (which had received a construction permit in 1973), Seabrook (which had received a construction permit in 1977), and other reactors stirred new interest in simplifying and streamlining the regulatory process. It seemed apparent that the complexity of the licensing process was a major deterrent to utilities who might consider building nuclear plants. By the late 1980s, the nuclear option looked more appealing to some observers, including some environmentalists, because of growing concern about the consequences of burning fossil fuel, especially acid rain and global warming. Furthermore, nuclear vendors were advancing new designs for plants that greatly reduced the chances of TMI-type and other severe accidents.

One way that the NRC proposed to facilitate licensing procedures was to replace the traditional two-step process with a one-step system to ease the burden on applicants. However, this raised a vitally important question: What level of detail would the NRC require in applications for advanced plants in order to satisfy its concerns about their safety? The agency had never required the detailed technical information in construction permit proposals that it expected in operating license applications. However, the NRC was uncertain as to how much data it would need in this one-step licensing process to evaluate and certify safety designs. After long discussions that reflected differing views among the Commissioners, the NRC staff, and nuclear vendors, the agency reached a decision on what constituted an "essentially complete design." It established a "graded approach" in which the level of detail that an applicant would be required to submit varied according to the relationship of the structures, systems, and components to plant safety. The objective of the NRC's action was to ensure safety while providing flexibility for the development of new designs.

Radiation Standards

While the NRC was deliberating over a number of new regulatory procedures and problems, it was also reviewing some old issues. The most prominent of those issues was radiation standards. The NRC had begun work on revising its radiation protection regulations in the aftermath of the TMI accident. Although the AEC had issued "design objectives" that in effect reduced the permissible levels of radioactive effluents from nuclear plants in the 1970s, the basic regulations for occupational and population exposure had remained unchanged since 1961 (an average of 5 rem per year for radiation workers and 0.5 rem annually for individuals in the general population). Based on new recommendations by NCRP and ICRP and on new research findings, the NRC tightened its regulations in several areas, the most prominent of which was to restrict population exposure to 100 millirem per year (rather than 500 millirem per year).

Despite new scientific information and epidemiological studies, the health effects of low-level radiation remained a source of uncertainty and controversy. Some studies provided results that were reassuring about the hazards of radiation emissions from nuclear plants. For example, a major survey conducted by the National Cancer Institute found no increased risk of cancer in 107 counties in the United States located near 62 nuclear power plants. However, other evidence was more disquieting, such as a cluster of cancer cases near the Pilgrim Nuclear Power Station in Massachusetts and a high incidence of leukemia in children around the Sellafield reprocessing plant in Great Britain.

None of the studies on the effects of low-level radiation were, or claimed to be, definitive. The subject continued to be a source of interest to, and debate among, scientists. It also continued to be a source of considerable anxiety to the public. The most graphic evidence of public apprehension about radiation was the public's reaction to the NRC's announcement of a new policy on radiation levels that were "below regulatory concern" (BRC). In June 1990, the NRC published a policy statement outlining its plans to establish rules and procedures by which small quantities of low-level radioactive materials could be exempt from regulatory controls. The agency proposed that if radioactive materials did not expose individuals to more than 1 millirem per year or a population group to more than 1,000 personrem per year, they could be eligible for the exemption. However, the NRC would not grant this exemption automatically; it would consider requests for exemptions for sites that met the dose criteria through its rulemaking or licensing processes. It intended that the BRC policy would apply to consumer products, landfills, and other sources of very low levels of radiation. The NRC explained that the BRC policy would enable it to devote more time and resources to major regulatory issues and thereby better protect public health and safety.

The NRC's announcement of its intentions on the BRC policy was greeted with a firestorm of protest from the public, Congress, the news media, and antinuclear activists. Some critics suggested that the agency was defaulting on its responsibility for public health and safety and that BRC policy would allow the nuclear industry to discard dangerously radioactive wastes in public trash dumps. One antinuclear group alleged that it was "a trade-off of people's lives in favor of the financial interests of the nuclear industry." In public meetings that the NRC held to explain BRC, aroused citizens called repeatedly for the resignation of the Commissioners or their indictment on criminal charges. Eventually, the Commission decided to defer any action on the BRC issue. The outcry over the BRC policy underscored the difficulty of even attempting to sponsor a calm and reasoned discussion on the subject of radiation hazards.

The uproar over the BRC policy was one of several indications of how the regulatory environment had changed since the passage of the Atomic Energy Act of 1954 made the development of nuclear power for electrical generation possible. A public that had welcomed the growth of nuclear power in the 1950s had become skeptical of the technology and suspicious of those responsible for its safety. Nuclear plants had become larger, more complicated, and more costly to build. Yankee Rowe Nuclear Power Station in Massachusetts, the longest running nuclear plant until its closure in 1992, had a capacity of 175 MWe, and its construction cost was about $39 million. For example, Seabrook by comparison had a capacity of 1,150 MWe and cost over $6 billion to build. The length and complexity of the licensing process had grown commensurately. The owners of Yankee Rowe applied for a construction permit in 1956 and received an operating license in 1960 without a murmur of protest. Seabrook's owners applied for a construction permit in 1973 and received an operating license in 1990 after long legal proceedings and many angry demonstrations. The contrasts between Yankee Rowe and Seabrook resulted from a series of interrelated technological, administrative, and political developments that shaped the history of nuclear regulation.

Chapter Four

New Issues, New Approaches

Chapter 4

The focus of the NRC's activities gradually shifted away from the licensing of new plants to overseeing the safety of operating plants. Because it received no applications for construction permits after 1978 and had completed work on most operating license applications a decade later, it devoted much less attention and fewer resources to its licensing responsibilities. During the first half of the 1980s, the NRC's deliberations and policy decisions were in large measure a response to the TMI accident. However, by the latter part of the decade, the agency was addressing a wide range of new questions related to the safety of the about 100 plants in operation. Not surprisingly, the issues that the agency considered often raised difficult and divisive questions for which there were no ready answers.

Decommissioning and License Renewal

One key issue related to operating plants that the NRC considered during the 1980s was decommissioning, the final step of the life cycle for licensed facilities. Between 1947 and 1975, a total of 50 nuclear plants, including five small experimental power reactors, were decommissioned. In the late 1970s, this experience gave the NRC confidence that the decommissioning of nuclear plants would not present major problems when their licenses expired. However, the NRC took a closer look at this subject in response to an investigation by the U.S. General Accounting Office, congressional hearings, and a petition from environmental organizations. In 1984, the staff reported to the Commission that existing regulations covered decommissioning in a "limited, vague, or inappropriate way and are not fully adequate." As a result, the NRC drafted a rule that required licensees to specify how they planned to ensure that sufficient funding was available to clean up the sites on which their plants were located and to make certain that radiation levels at decommissioned sites were low enough to allow the land to be used for other purposes. After soliciting public comments and making modest revisions in the draft, the NRC published a final rule in 1988.

The decommissioning rule was much more comprehensive than earlier NRC regulations, but it did not resolve all of the issues that arose on the subject. Within a short time after the rule became final, the agency faced an unprecedented and unanticipated question: What should be done about funding for "prematurely shutdown reactors"? The closing of three plants, including Shoreham, well before their operating licenses expired raised questions about how to pay for the costs of decommissioning reactors that had not operated long enough to accumulate adequate funding. This issue was underscored by the fact that the costs for decommissioning the Yankee Rowe plant ran much higher than projected. While the NRC wrestled with this question, it also deliberated over the level of radiation that should be permitted at the sites of decommissioned plants. This issue generated opposing views and sometimes sharp differences between the NRC and the U.S. Environmental Protection Agency.

As decommissioning issues were debated, the NRC also devoted considerable attention and resources to renewing the operating licenses of nuclear power plants. Although some utilities were closing reactors long before the expiration of their 40-year operating licenses, others were weighing the possibility of extending the lives of plants beyond 40 years. The 40year licensing period for nuclear plants was a rather arbitrary compromise written into the Atomic Energy Act of 1954. It was not based on technical information or operating experience but instead on the amortization period for fossil fuel plants. In the late 1970s, industry groups closely examined the issue of plant life extension for the first time. For example, the Electric Power Research Institute concluded that the reconditioning of old plants offered potentially major benefits, but it cautioned that the benefits depended on financial considerations and on technical assessments, environmental issues, and projections of power availability. Those uncertainties were compounded by industry's concern that the NRC was not prepared to address the question of license renewal promptly and knowledgeably.

In 1985, Chairman Nunzio J. Palladino prodded the NRC to undertake a careful analysis of license renewal. The agency had sponsored research on the critical question of the safety effects of plant aging for years, but many technical questions remained unanswered. License renewal also raised complex legal and policy issues. The NRC staff cited the "central regulatory question" that plant life extension presented: What is an adequate licensing basis for renewing the operating license of a nuclear power plant?

The NRC deliberated over this issue and its corollaries for several years. Eventually, it decided that the maximum length of an extended license would be 20 years. The agency also concluded that using the existing regulatory requirements governing a plant would offer reasonable assurance of adequate protection if its license were renewed, provided that the "current licensing basis" was modified to account for age-related safety issues. In 1991, the Commission approved a regulation on the technical requirements for license renewal. After considering ways to evaluate the environmental consequences of license renewal, the NRC elected to develop a generic environmental impact statement that covered effects that were common to all or most nuclear plants. In April 1998, Baltimore Gas and Electric Company became the first utility to apply for license renewal for its Calvert Cliffs plants on the Chesapeake Bay. Duke Energy Corporation followed suit in July 1998 when it sought license extensions for the Oconee Nuclear Station in South Carolina.

Risk Assessment and Nuclear Safety

As the NRC considered its policies on license renewal, representatives of the nuclear industry expressed concern that the costs and uncertainties of the regulatory process would negate the potential advantages of plant life extension. This concern was consistent with strong industry criticism of the NRC's regulations or the ways in which these regulations were implemented. Of course, industry protests about regulatory burdens were nothing new, but they had taken on increased urgency and intensity by the early part of the 1990s. Industry officials complained that NRC regulations were in many cases excessive and potentially counterproductive. They particularly objected to the agency's numerical ratings of plant performance, which they found to be arbitrary and inconsistent. They also asserted that many of the requirements imposed in response to the TMI accident gave the NRC an unduly intrusive presence in the day-to-day operations of its licensees.

A report prepared by the Towers Perrin consulting firm for a prominent industry group, the Nuclear Energy Institute (NEI), concluded in 1994 that the NRC's policies and practices represented a "serious threat to America's nuclear energy resource" by distracting plant management, undermining public trust in nuclear power, and "pricing nuclear power out of the competitive energy marketplace." The report called for prompt changes to "reverse the NRC's role in accelerating the decline of the nuclear industry." The Towers Perrin study found that the NRC regulatory approach was "negative and punitive," and it urged the agency to place greater emphasis on performance-based assessments that would recognize the significant improvements that industry had achieved since the TMI accident.

By the time that the Towers Perrin report appeared, the NRC had begun to evaluate ways in which risk assessment and performance indicators should be factored into the regulatory process. Nuclear industry representatives complained that the NRC relied too heavily on "prescriptive" regulations that specified a rigid solution to a licensee on how to carry out a safety goal. In some cases, this meant that licensees that had already met a regulatory standard using their own methods had to expend considerable staff hours to implement an alternative approach that the NRC preferred. The Towers Perrin report urged the NRC to place greater emphasis on nonprescriptive performance-based regulations. This would allow licensees greater leeway in determining how to accomplish regulatory goals and presumably cut costs without sacrificing safety. Noting the rise in operating and maintenance costs, NRC Chairman Ivan Selin (1992–1995) declared, "We feel that the NRC has been a factor in this and that perhaps it's time for us to step up our search for places where we may inadvertently cause more costs than justified by health and safety." In 1991, the Commission instructed the NRC staff to investigate the feasibility of using more performance-based regulations that focused on a "result to be obtained, rather than prescribing to the licensee how the objective is to be obtained." This initiative received strong support from Selin; his successor, Shirley Ann Jackson (1995–1999); and their colleagues on the Commission.

The effective employment of performance-based regulations was closely tied to informed analyses of risk. The Towers Perrin report complained that the NRC made little effort to distinguish safety from nonsafety issues and to appropriately prioritize them. It claimed that the result was a "diversion and dilution of licensee resources" away from the most important safety issues, such as human performance when problems arose at the plant. The industry and many in the NRC called for the conduct of probabilistic risk assessments (PRAs) as a more effective way to assess hazards and to use resources efficiently to protect against them.

The benefits of PRAs had been debated within the NRC since the Rasmussen report of 1975 without making a major impact on the formulation or enforcement of the agency's regulations. Industry was concerned that the NRC remained wedded to a "deterministic analysis" and a redundant "defense-in-depth" approach that downplayed the role of risk assessment in safety evaluations. Regulators using a deterministic approach simply tried to imagine "credible" mishaps and their consequences at a nuclear facility and then required the defense-in-depth approach—layers of redundant safety features—to guard against them. Before TMI, no severe accidents that melted the core of a plant had ever occurred, and no sure way existed to calculate the probability of a major accident. NRC experts used their collective judgment to determine what accidents were credible, and the agency often mandated multiple safety systems to compensate for the uncertainty of an accident's probability and consequences. This approach had worked well in protecting public safety; defense-in-depth was critical in preventing sizeable releases of the most dangerous forms of radiation at TMI. However, the defenseindepth approach was not effective in prioritizing accidents or in judging when an extra, often expensive, safety system produced a commensurate increase in margins of safety. Proponents argued that a PRA, with its much more detailed use of probabilities and modeling of plant and human behavior, could better deal with such issues.

During the 1980s, the NRC moved cautiously to implement PRAs. Industry and the NRC agreed on the general objective of increasing the use of PRAs, but many uncertainties about how to apply the concept of risk assessment in practice remained. Too few data were available to allow PRAs to offer reliable estimates of risk. For example, an NRC tabulation of PRA data on the probability of core melting indicated that uncertainties with the data and models meant that the actual risk could be higher or lower by an order of magnitude. A U.S. General Accounting Office report of 1985 praised the NRC's decision to limit the applicability of PRAs to providing supplemental information in environmental impact statements, helping to prioritize the resolution of safety issues that were generic to all reactors, and determining the possible benefit of proposed regulatory actions. The U.S. General Accounting Office concluded that "substantial limitations of PRA in terms of the uncertainties of the results" supported the continuation of its qualified use in safety analysis and decisionmaking.

By the 1990s, the state of the art of PRAs had advanced significantly, in part as a result of research programs launched in the wake of the TMI accident. In 1991, the NRC completed a major study entitled, "Severe Accident Risks: An Assessment of Five U.S. Nuclear Power Plants," that it hailed as a "significant turning point in the use of risk-based concepts." In 1995, the Commission unanimously adopted a policy statement that encouraged the broad application of PRAs in the regulatory process to enhance decisionmaking on safety issues and to ease "unnecessary burdens on licensees." Within a short time, the agency began to use the phrase "risk-informed, performance-based regulation" to describe its intention to take advantage of the insights provided by risk assessment. The NRC suggested that risk analysis would enable it to "focus on those regulated activities that pose the greatest risk to the public." Nevertheless, the policy statement made clear that PRAs were still playing second fiddle to the defense-in-depth approach and should be used largely to identify "overly conservative regulatory requirements." The continuing precedence of deterministic analysis was highlighted in 1997 when the Commission voted to require a containment spray system in a new Westinghouse plant design even though PRAs indicated that the design was "safe enough" without the spray system.

Despite the affirmation of the importance of defense in depth, the NRC continued to search for ways to use PRAs to improve the regulatory process. Eventually, it developed a "reactor oversight process" to "inspect, measure, and assess the safety performance of commercial nuclear power plants and to respond to any decline in performance." The NRC evaluated individual plants on a series of performance indicators with regard to reactor safety, radiation exposures to workers and the public, and physical protection.

Quality Assurance and Plant Maintenance

One of the most important issues that the NRC tackled as it turned its attention to the regulation of operating reactors during the 1980s was quality assurance (QA), which had been an issue of growing concern since the waning days of the AEC. In 1974, John G. Davis, a regulatory staff member in the AEC, told attendees at a nuclear industry meeting that "considering the extent that AEC has gone in order to stress the importance of QA, we find the continuing deficient programs to be quite disappointing." He suggested that utilities, rather than viewing QA as an essential part of plant management, too often merely met the minimum requirements mandated by the AEC. To improve existing practices, in addition to providing a "higher incentive" for performance through stiffer fines for failures, the AEC introduced a trial program of "resident inspectors" at two plant sites. Their assignment was to provide regular onsite verification that utilities complied with regulations instead of relying on comparatively superficial, infrequent visits from inspectors based in regional offices. In 1977, the NRC determined that the resident inspector concept was workable and expanded the program to more facilities.

In the wake of the TMI accident, the NRC gave increased attention and resources to QA and proper maintenance in operating plants. The agency estimated in 1985 that more than 35 percent of the "abnormal occurrences" that it had reported to Congress over the previous 10 years were directly attributable to maintenance deficiencies. Many of the problems arose from human errors, such as failing to follow procedures, installing equipment incorrectly, or using the wrong parts to make repairs. The need for improvements in maintenance was underscored when an incident in 1985 at the Davis-Besse Nuclear Power Station in Ohio resulted in the loss of all feedwater. Failures in feedwater pumps, including auxiliary pumps that the plant had not tested or maintained, caused what could have produced a major accident.

The nuclear industry was well aware of its shortcomings in maintenance programs and took steps to make improvements. The NRC applauded those efforts but concluded that the licensees still "had a long way to go in the maintenance area." Therefore, in June 1988, the Commission directed the NRC staff to draft a maintenance rule as a matter of "highest priority." In June 1991, despite industry objections that the rule was unnecessary, the Commission voted to adopt a regulation that required adequate maintenance programs in all commercial nuclear plants. It ordered the staff to prepare broad guidelines to assist licensees in identifying existing weaknesses and in establishing procedures that would fulfill the NRC's requirements.

The maintenance issue was an important part of the larger problem of QA. After the TMI accident, the NRC decided to station two resident inspectors at each plant. As the number of inspectors and inspections increased, the hours devoted to checking on licensees' performance doubled by the early 1980s. At the same time, the number of inspection procedures more than doubled, which made it difficult for the resident inspectors to complete their tasks on schedule. As a result, the NRC began to make greater use of risk assessment and trend analysis in its overall QA and inspection protocols. Risk assessment presented the possibility of prioritizing those inspection activities most relevant to safety and reducing the regulatory burden on licensees.

In 1987, the NRC staff announced a shift to performance-based inspections. The staff would use the direct observation of plant activities for the purpose of enhancing safety and reliability instead of document reviews that simply demonstrated that a licensee conformed to regulations and procedures. However, implementing performance-based inspections proved difficult. The Towers Perrin report testified to the industry's view that inspectors' evaluations were frequently inconsistent and arbitrary. In 1995, the NRC's inspector general concluded that the agency's resident inspectors lacked a clear understanding of how to carry out the performance-based concept. This conclusion led to a revision of inspection guidelines and reforms in the training of inspectors.

The NRC improved its use of data and risk assessment in inspections. In 1995, the South Texas Nuclear Power Plant suffered an extended shutdown after continuing problems with emergency systems. The NRC found that many of the plant's problems were evident from earlier inspection reports, but that the information had not been applied effectively in overall performance assessments. The agency sought to make better use of inspection data and reformed its inspection program through greater emphasis on PRA. In 2002, for example, it issued a new Standard Review Plan as guidance for the use of PRAs in the inservice inspection of piping.

The Millstone Controversy

Although risk-informed regulation offered many potential benefits for evaluating the performance of nuclear plants, it was not capable of detecting every safety issue that could generate acute public concern. In that regard, risk-informed regulation was not necessarily a useful means of building public confidence in nuclear power technology or in the NRC. This fact was amply demonstrated when a series of problems arose at the Millstone Power Station, which included three plants located on the northern side of Long Island Sound in Connecticut. The safety issues at Millstone required attention, but they were not so serious that risk analysis was likely to identify them as priority matters. Commissioner Nils J. Diaz commented in 1997 that of the many issues to be resolved at Millstone, "only a handful appear to have been safety-significant." Nevertheless, the failures at Millstone created a great deal of controversy and a barrage of criticism of the NRC.

The uproar over Millstone began in the early 1990s when several plant employees claimed that they were harassed, intimidated, or dismissed from their jobs by the owner of the plant, Northeast Utilities, for calling attention to safety problems and violations of NRC regulations. The NRC investigated the concerns raised by these "whistleblowers" and determined that the safety issues that they raised were not of major significance and had been corrected. However, the agency also concluded that the utility had harassed employees and assessed a fine against it of $100,000, the maximum amount allowed by law. The NRC's action in this case did not satisfy the dissidents at Millstone or elsewhere, who insisted that the agency was neither prompt nor firm in dealing with the issues that they cited or in protecting them from retaliation by their employers. As a result of the complaints from Millstone and other plants, the agency reexamined and eventually tightened its policies to better protect whistleblowers who contacted the NRC about safety issues.

Meanwhile, new revelations at Millstone generated increasing NRC scrutiny. They also commanded growing media attention, much of which was sharply critical of the NRC. In 1993 and again in 1994, the NRC fined Northeast Utilities for procedural violations that the agency viewed as serious lapses in the management of the Millstone units. The utility pledged to improve its performance and "to resolve issues raised by [its] employees." Nevertheless, another issue that company employees reported soon triggered new reservations about safety at Millstone and the effectiveness of the NRC's enforcement policies. In this case, the whistleblowers objected to the company's practice of placing the entire nuclear core into the spent fuel pool at Millstone, Unit 1, during refueling operations. NRC regulations specified that in older plants such as Millstone, Unit 1, only one-third of the spent fuel rods could be moved into the pool. However, Millstone, Unit 1, had performed "full-core offloading" for years as an "emergency" procedure, with the NRC's knowledge of this practice. Finally, after its employees questioned the practice, Northeast Utilities applied for a license amendment that expressly permitted full-core offloading, and the NRC granted its approval in November 1995.

By that time, the utility and the NRC were the subjects of extensive and unflattering coverage in the local media. In March 1996, the criticism reached a new level of visibility when Time magazine ran a cover story on the whistleblowers who had "caught the Nuclear Regulatory Commission at a dangerous game." The article suggested that an accident in a spent fuel pool posed the hazard of "releasing massive amounts of radiation and rendering hundreds of square miles uninhabitable." It charged that the NRC "may be more concerned with propping up an embattled, economically strained industry than with ensuring public safety." NRC Chairman Jackson conceded that the Time article demonstrated that "not all aspects of nuclear regulation or nuclear operations in certain places are as they should be," but she strongly denied the implication that "the Millstone situation borders on an impending TMI- or Chernobyl-type disaster."

Amid the growing criticism, the NRC conducted its own reviews to identify and correct errors that the Millstone experience brought to light. An internal task force reported in September 1996 that the "safety significance of Millstone's refueling practices was low." Nevertheless, it recommended a series of procedural, informational, and management improvements designed to ensure that licensees complied with NRC regulations and that the agency enforced its own rules.

The NRC sought to minimize "recurring exemptions" from its regulations, such as those that occurred in the refueling practices at Millstone, Unit 1. It reemphasized its position that exemptions were intended to apply to special circumstances in which specific requirements could be waived without compromising public safety. The agency also undertook a careful study of a frequently used provision in its regulations that allowed licensees to make changes in their plants without NRC permission under certain conditions. In 1999, the Commission approved revisions designed to clarify the rule and provide guidance on when NRC consent was necessary within a risk-informed framework.

While the NRC examined its own regulations and procedures, it conducted an expanding probe of the Millstone plant. In May 1996, the NRC's inspector general faulted the agency for failing to recognize the problems at Millstone and impose corrective actions much earlier. When the NRC's investigations, along with those conducted by the utility, turned up hundreds of performance and procedural deficiencies, the agency took the unusual step of stipulating that the utility would not be allowed to restart its three units, all of which had been shut down, without a formal vote of the Commission. Eventually, after the utility made management changes, took a series of steps to address its problems, and decided to permanently close Millstone, Unit 1, the Commission authorized the restart of Unit 2 (in 1999) and Unit 3 (in 1998). The series of problems at Millstone threw into sharp relief the general difficulties that the NRC had encountered with plants that did not perform up to standards and did not correct their deficiencies promptly or effectively. The Commission devoted a great deal of energy to encouraging or forcing improvements in plants that did not fully meet its requirements.

Regulating Nuclear Materials

Although reactor safety issues captured a lion's share of public notice, the NRC also devoted substantial resources to a variety of complex matters in the area of nuclear materials safety and safeguards. The protection of nuclear materials from theft and diversion remained a major agency concern, although it did not command the level of public attention it had received during the 1970s. In cooperation and sometimes in conflict with other Government agencies, the NRC evaluated the safety problems involved in building and operating repositories for high- and low-level radioactive waste. Despite Federal legislation that attempted to provide the means for establishing permanent waste sites and the efforts of Government officials, scientists, engineers, and other professionals, the disposal of radioactive wastes remained a source of intense public concern and bitter political controversy. The NRC also considered its role in regulating certain medical uses of radioactive materials. Although it exercised only limited responsibilities in the field of "radiation medicine," it sought to ensure that patients received the proper doses of radiation from procedures under its regulatory authority. The agency's rules elicited protests from medical practitioners and organizations who complained about regulatory overkill that intruded into physician-patient relationships.

The issues surrounding the regulation of nuclear materials, the problems at Millstone, and the use of risk assessment in regulatory decisionmaking underscored the prevailing patterns in the history of nuclear regulation over a period of four decades. The nuclear industry and materials licensees often asserted that regulatory requirements were too burdensome, too inflexible, and too strict. On the other hand, nuclear critics frequently lamented that regulatory requirements were too lax, too sympathetic to industry concerns, and too inattentive to public safety. The NRC, and the AEC before it, attempted to find a proper balance between essential and excessive regulation, but this task was difficult and uncertain, and it usually elicited complaints from one side or all sides of regulatory issues. The NRC sought to separate valid criticisms from those that were exaggerated or ill formed, but this process received little praise from the agency's different (and frequently competing) constituencies. "The bane of the regulator," a senior agency official remarked in 1998, "is to feel unloved." The ongoing effort to promote the safe operation of nuclear power plants without imposing undue burdens on their owners ensured that nuclear regulation would remain a complex and controversial public policy issue.

Chapter Five

A Terrorist Attack and a Nuclear Revival

Chapter 5

Shortly after the dawn of the 21st century, the NRC marked the 25th anniversary of the date that it began operations in January 1975. It continued to face many of the same issues and controversies that had been a prominent part of the first quarter of a century of its history. However, the new century soon brought unexpected and unfamiliar new developments that had a major impact on the agency's policies, procedures, and planning for the future.

The Impact of the Terrorist Attacks of September 11, 2001

The first such event was the shock of the terrorist attacks on the World Trade Center in New York and the Pentagon building near Washington, DC, on September 11, 2001. The air assaults by suicide squads raised two crucial questions for the NRC and the nuclear industry: (1) the vulnerability of nuclear plants to a raid by terrorists who could disable safety systems and cause a massive release of radiation to the environment and (2) the possible effects of an airplane loaded with fuel hitting a plant at a high speed.

As soon as the NRC learned of the attacks on the morning of September 11, 2001, it told its licensees to move to their highest level of readiness, Security Level 3. This meant that licensees added to the size of their security forces on site, increased the number of patrols that they conducted, and made access to plants more difficult. The NRC pointed out that security arrangements at nuclear plants before the September 11 attacks had already been rigorous as a result of the regulations that had been imposed during the 1970s. In September 2002, Chairman Richard A. Meserve commented, "I am aware of no other industry that has had to satisfy the tough security requirements that the NRC has had in place for a quarter of a century."

In the aftermath of the September 11 attacks, the NRC reviewed its regulations to consider what steps should be taken to improve existing requirements. In February 2002, it ordered a series of security measures, many of which formalized steps taken immediately after the terrorist attacks. For security reasons, the NRC offered few details to the public about the enhanced requirements. However, it generally instructed licensees to conduct more patrols, build up the size and capability of security forces, install additional physical barriers, perform vehicle inspections at a greater distance from reactors, provide tighter control of access by plant workers to buildings and equipment, and improve coordination with military and law enforcement agencies. The NRC also decided that each plant would carry out force-on-force exercises to evaluate the effectiveness of its security regime every 3 years instead of every 8 years. In April 2002, the Commission created the Office of Nuclear Security and Incident Response to serve as the focal point for the NRC's security programs. In April 2003 and March 2006, the NRC issued upgraded requirements for the "design-basis threat" that plant owners had to be prepared to meet. The provisions were not made public, but the agency announced that these requirements would guard against "the largest reasonable threat against which a regulated private guard force should be expected to defend under existing law."

The regulatory changes that the NRC made in the wake of the September 11 attacks stirred criticism from those who believed that nuclear plants were still vulnerable. Some members of Congress, claiming that the NRC and industry had failed to adequately address the dangers of terrorist attacks, introduced legislation to establish a Federal guard force under NRC authority. The NRC strongly objected to the proposal on the grounds that it would be "a costly, unwieldy solution" that would not benefit security but would compromise the agency's ability to promote reactor safety. In 2003, Daniel Hirsch of the Committee to Bridge the Gap, David Lochbaum of UCS, and Edwin Lyman of the Nuclear Control Institute accused the NRC of keeping a "dirty little secret"—that it required nuclear plant owners "to maintain only a minimal security capability." They asserted that the defensive posture envisioned by the design-basis threat would leave security forces ill prepared and ill equipped to fight off a well-armed band of commandos who were intent on gaining access to a plant and causing a massive release of radiation. They further contended that the simulated attacks that the NRC used to test a plant's readiness, called Operational Safeguards Response Evaluation (OSRE), showed severe weaknesses. "At nearly half the nuclear plants where security has been OSRE-tested," they wrote, "mock attackers have been able to enter quickly and simulate the destruction of enough safety equipment to cause a meltdown."

The NRC strongly disagreed with the charges that its security requirements were lax and ineffective. Commissioner Edward McGaffigan, Jr., was particularly outspoken in responding to such indictments. He denied suggestions that nuclear plants were "soft targets" and emphasized that they were "hard targets by any conceivable definition." He accused critics of distorting the results of the OSRE drills. "These were not pass-fail exams," McGaffigan remarked. "They were meant to identify weaknesses that needed to be corrected." He pointed out that although the mock assault teams had "almost perfect knowledge of the plant's defenses and perfect knowledge of the plant's layout and the equipment they need to attack to try to bring about core damage," they succeeded in reaching their targets in only 9 of the 59 exercises carried out between April 2000 and August 2001. In addition, the successes that they achieved revealed flaws that were "promptly fixed." The NRC made further improvements in the program after the September 11 attacks. The security of plants from a ground attack continued to be a source of controversy and reevaluation. For example, in September 2003, the U.S. General Accounting Office reported that despite the actions taken after the September 11 attacks, the NRC needed to improve the collection and dissemination of information, tighten inspection and access procedures, and plan more realistic exercises. In November 2004, the NRC began to carry out drills that reflected improvements made after the September 11 attacks, including the new design-basis threat and more realistic scenarios.

As the NRC was working on the protection of plants from a commando strike, it was also considering another problem that was equally difficult and even more ethereal—the effects of an airplane hitting a reactor building or spent fuel pool. Shortly after terrorists flew airplanes into the World Trade Center and the Pentagon on September 11, 2001, the NRC acknowledged that nuclear plant builders "did not specifically contemplate attacks by aircraft such as Boeing 757s and 767s, and nuclear plants were not designed to withstand such crashes." The only operating plant designed to guard against the impact of a large airplane was TMI, located 3 miles from Harrisburg International Airport. It was designed to protect against a plane of about 200,000 pounds accidentally hitting the plant at a speed of 230 miles per hour; the planes that terrorists hijacked on September 11, 2001, were heavier and hit their targets at speeds of 350 to 537 miles per hour. Although the NRC pointed out that containment buildings were "extremely rugged structures," it could not predict with certainty what the consequences would be "if a large airliner, fully loaded with jet fuel...crashed into a nuclear power plant." The critical issue that industry and the NRC then faced was to assess the vulnerability of plants to an air attack that could produce a massive release of radiation.

In June 2002, NEI announced the results of a study conducted by the Electric Power Research Institute that it had sponsored on this issue. "We think it's extremely unlikely that the aircraft would be able to penetrate the reactor," an NEI official declared. "We feel very, very confident about the containment structure." The report analyzed the effects of a plane hitting the reactor building at various angles at about 350 miles per hour. It did not consider the impact of a plane traveling at a greater speed because the probability that a pilot could strike the target at a high speed and at low altitude was "virtually nil." Nuclear critics were not convinced. Lyman questioned the methodology of the NEI study and contended that an airplane piloted by a terrorist could indeed crash through containment with catastrophic consequences. The NRC, based on the research of national laboratories and its own staff, arrived at conclusions that were supportive of, but more equivocal than, those of NEI. In September 2004, the agency reported that if an airplane struck a nuclear plant, it could cause radiation releases. However, the NRC found it "unlikely" that a crash would lead to a large release of radioactive materials and emphasized that plant operators would have sufficient time to take "mitigating actions" to protect public health.

Nuclear critics argued that even if the containment structure is strong enough to withstand the impact of an airplane, spent fuel pools are much more vulnerable. The pools that hold highly radioactive fuel rods, after their removal from the core, are housed in separate buildings that are not as robust as the containment structures that protect reactors. The fuel rods are stored under at least 20 feet of water, which is enough of a barrier to prevent radiation exposure to persons standing above the pools. The walls of the pools are built with steel-reinforced concrete that is 4 to 6 feet thick. In 2003, a group of eight respected nuclear critics published an article that claimed that a terrorist attack with an airplane or an antitank missile could drain the cooling water from a spent fuel pool, ignite a large fire, and cause consequences "significantly worse than those from Chernobyl." The article was often referred to as the "Alvarez report" after the first-listed author, Robert Alvarez of the Institute for Policy Studies.

The NRC staff carefully reviewed the Alvarez study. It concluded that the report suffered from "excessive conservatisms" and failed to make its case for the need for costly measures to improve the security of spent fuel storage. Alvarez and his coauthors had drawn heavily from earlier studies that the NRC had conducted or sponsored, and the agency commented that most of these studies "are not applicable to terrorist attacks." It revealed that research performed since the September 11 attacks showed that the hazards cited in the Alvarez report were overstated and misleading and that existing methods of storing spent fuel were sufficient "to adequately protect the public." Alvarez and his colleagues complained that the NRC had criticized their findings but had refused to make public the new classified studies on which it based its position. They accused the NRC of hiding its analysis "behind a curtain of secrecy."

The standoff between the NRC and its critics on the vulnerability of spent fuel pools led Congress to request a study of the issue by the National Academy of Sciences. A group of 15 scientists conducted the investigation and announced their findings in April 2005. The group concluded that a successful terrorist attack would be difficult to execute but would be possible under some conditions. The panel argued that there were "no requirements in place to defend against the kinds of larger-scale, premeditated, skillful attacks that were carried out on September 11, 2001." The NRC announced that it "respectfully" disagreed with that contention. It also suggested that even if a spent fuel pool were drained, a fire hose or two could provide enough water to cool the fuel rods. The philosophical differences between the National Academy and the NRC were not easily resolved because the agency, in accordance with legal requirements, could not share sensitive, although unclassified, information about defensive measures at nuclear plants.

This issue led to sharp exchanges with the National Academy, which complained that guidelines for making "safeguards information" available were vague. The NRC suffered stinging criticism for its position. For example, the New York Times found it "disturbing that the commission, in the name of national security, denied the academy the information needed to assess the effectiveness of security improvements instituted since 9/11." It called the dispute a "sorry episode." Despite the acrimony of the debate, the NRC carried out one of the National Academy's major recommendations by instructing licensees to reposition fuel rods in spent fuel pools in a way that would reduce the buildup of heat and decrease the chances of a disastrous fire.

"Significant Degradation" at Davis-Besse

At the same time that the NRC was evaluating security requirements at nuclear power sites, it was responding to a serious safety issue that arose at the Davis-Besse plant in Ohio. In February 2002, an inspection of the reactor revealed "significant degradation" of the pressure vessel lid. To the surprise and consternation of the company that owned the plant, First Energy Nuclear Operating Company, and the NRC, it turned out that corrosion had created a "large cavity" in the vessel head. The football-sized gap measured about 4 inches wide and 7 inches deep. The corrosion had displaced about 70 pounds of steel and left only a comparatively thin layer of stainless-steel cladding about three-eighths of an inch in depth. The damage to the head was very disturbing because a failure of the corrosion-resistant cladding could have led to a loss-of-coolant accident.

The discovery of the corrosion of the reactor vessel head raised a number of troubling questions. The critical issue was why the utility and the NRC had failed to identify the problem sooner and take action to correct the conditions that caused the damage. Investigations by First Energy and the NRC revealed that the company had paid insufficient attention to signs of corrosion and had made erroneous assumptions, based on incomplete information, about the need for careful inspection of the head. The utility found that "there was a focus on production, established by management, combined with taking minimum actions to meet regulatory requirements, that resulted in the acceptance of degraded conditions." Lew Myers, Chief Operating Officer of First Energy, told the NRC that he was "humbled and in fact embarrassed" by those findings.

The NRC established a "Lessons Learned Task Force" to examine the agency's role in the deficiencies at Davis-Besse. It concluded that staff shortages and the attention commanded by other troubled plants in the NRC's Region III office had contributed to the delay in finding the corrosion at Davis-Besse. Davis-Besse was regarded as a "good performer," and the regional office focused its resources on other plants that were shut down and that required augmented oversight. The number of inspection hours at Davis-Besse was consistently below average for the region, and job openings for resident inspectors at the facility went unfilled for lengthy periods. The task force also criticized the performance of the resident inspectors and faulted them for not recognizing the severity of the corrosion problem, reporting it to superiors, or following established procedures for dealing with it.

The failures at Davis-Besse generated a great deal of concern within First Energy and the NRC. The issue received considerable coverage in newspapers around the country and extensive treatment in Cleveland, Toledo, and other locations in Ohio. The problem of corrosion at Davis-Besse soon became linked to another controversy over the NRC's inquiry into a generic problem of potential cracking in control rod drive mechanism nozzles in the vessel head. On August 3, 2001, the NRC instructed owners of pressurized-water reactors to check the status of the drive mechanism nozzles in their plants by December 31, 2001. The agency acted in

The discovery of extensive corrosion on Davis-Besse's reactor vessel head led to an extensive shutdown for the plant's utility company and considerable reform of NRC inspection practices

response to the discovery of "circumferential cracking" at two pressurized-water reactors, a defect that could over time cause a serious accident. The inspections would have to be performed when plants were shut down for refueling or other reasons, and the NRC specified that the date for conducting the surveys could be moved back if the staff judged on a case-by-case basis that the risks were acceptably small. First Energy petitioned the NRC to postpone the Davis-Besse inspection until a scheduled outage on March 31, 2002. The NRC staff reviewed the request and determined that the utility could run the plant until February 16, 2002, without triggering a "significant safety concern."

The question of the timing for inspecting the drive mechanism nozzles for cracking soon generated an intense debate within the NRC. First Energy had unexpectedly discovered the corrosion of the vessel head in February 2002 in the process of looking for evidence of damage to the nozzles. This discovery threw into sharp relief the issue of whether the NRC had erred in allowing the plant to operate for 6 weeks beyond the December 31, 2001, deadline for inspecting the nozzles. In December 2002, the NRC's inspector general sent a report on Davis-Besse to the Commission. The inspector general had undertaken the study in response to charges from UCS that the agency had failed to adequately regulate the plant and that a loss-of-coolant accident could have been the result of this failure. In the report, the inspector general strongly criticized the NRC's performance. It found, among other things, that the agency had "considered the financial impact to the licensee of an unscheduled plant shutdown" rather than making public health and safety its highest priority.

In an unusually unvarnished response, the NRC Commissioners told the inspector general that although they agreed with some aspects of the report, they regarded the most serious criticisms as "unjustified, unfair, and misleading." They were especially incensed by the suggestion that they had placed the financial well-being of First Energy above public health and safety. They pointed out that the NRC had permitted the short extension beyond the original deadline for inspection of the reactor vessel head at Davis-Besse only after careful consideration by the staff. Furthermore, the Commissioners admonished the inspector general for not anticipating that the report "would be misconstrued to suggest staff acceptance of the unexpected head corrosion at the Davis-Besse plant." They complained that the "staff did not know about the head corrosion at the time of its decision and, quite frankly, it is Monday-morning quarterbacking to question the decision on [circumferential] cracking in the false light of subsequent knowledge."

The Commissioners' concern that the original problem of the cracking of the control rod drive mechanism nozzles would be confused with the more urgent and more alarming problem of corrosion of the reactor head was well founded. The inspector general's report provoked a barrage of attacks on the NRC, at least some of which were based on the erroneous premise that the agency had authorized Davis-Besse to continue operation even though it knew about the corrosion of the reactor head. The Toledo Blade reported that the NRC had shown "reckless complacency" by coming "down on the side of corporate profits" in a way that led to a "near calamity at the plant." The Plain Dealer denounced NRC Chairman Meserve and his "equally narrow-minded" colleagues for "badmouthing" the inspector general's report and charged that they had failed "to put safety at the top of their agenda."

The controversy over Davis-Besse continued even after First Energy completed the repairs on the vessel head and the NRC allowed it to resume operation in March 2004. The central question concerned the possible consequences had the stainless steel cladding on the inside surface of the head failed. Critics of the NRC and First Energy claimed that the plant was on the verge of a catastrophic accident. Paul Gunter of the Nuclear Resource and Information Service accused the NRC of obscuring "just how close we were to losing Toledo." The NRC readily conceded that a break in the cladding could have led to a loss-of-coolant accident and that the corrosion of the head was an "enormous failure" on the part of the agency and the utility. However, it denied that fracture of the cladding would have inevitably led to a massive release of radiation from the plant. The agency emphasized that the other barriers (including the containment building) that are in place would have, in all likelihood, prevented the escape of radiation. NRC Chairman Nils Diaz pointed out that "a variety of safety systems was available" and that even if the stainless steel liner "had been breached, the layers of safety would have protected Ohioans."

A Nuclear Revival?

Even as the NRC was dealing with new challenges to reactor safety from terrorist attacks and from the lapses at Davis-Besse, the nuclear power industry was showing its first signs of revival after a slump of more than two decades. In the aftermath of the TMI accident, the nuclear industry adopted a series of reforms to correct deficiencies that the accident had so graphically revealed. Changes in operator training, plant management, control room design, and equipment led to significant improvements in the safety and reliability of nuclear power. The capacity factor for nuclear plants indicates the percentage of time during which the plants produce power. This factor increased from 50 to 60 percent in the 1970s to 90 percent currently. The cost of generating electricity from nuclear reactors fell significantly. A series of safety indicators, including the number of reactor scrams (the sudden shutting down of a nuclear reactor), safety system failures, and collective radiation exposure for plant workers, showed consistent and substantial industrywide improvement. Nevertheless, the long-term prospects for the nuclear industry did not look promising. During the 1990s, it appeared doubtful that any new reactors would be built because of the high capital costs of construction relative to other sources of power. "The industry is doing better now," Matthew Wald wrote in the New York Times in March 1999, "but ironically extinction is in sight."

During the early years of the 21st century, however, the outlook for nuclear power brightened considerably. One important reason was the increasing need for power. During the 1990s, energy consumption in the United State grew by about 23 percent while production expanded by less than 3 percent. It seemed apparent that many new plants would have to be built to generate enough power to meet America's energy demands. In May 2001, President George W. Bush's administration estimated that the Nation would need at least 1,300 and perhaps 1,900 new power plants over a period of two decades. The disadvantages of fossil fuel as a source of energy were evident. A growing percentage of the country's oil came from politically unreliable nations and domestic re-fining capacity had declined substantially. Coal was plentiful but exceptionally dirty. Natural gas had been the fuel source of choice during the 1990s, but there were acute concerns about the adequacy of supplies and cost. In that context, nuclear power began to look attractive or at least worthy of consideration. "We have even seen the first stirring of interest in the possibility of [nuclear plant] construction in the United States—a thought that would have been unthinkable even a year ago," commented NRC Chairman Meserve in March 2001.

The advantages of nuclear power appeared even more apparent, at least to some observers, as public knowledge and concerns about global warming grew rapidly in the first decade of the 21st century. Scientists had theorized about the possibility that increasing quantities of carbon dioxide in the atmosphere could lead to climate change as far back as 1899. In 1965, a report to the President's Science Advisory Committee suggested that increasing levels of carbon dioxide from fossil fuels, which it called the "invisible pollutant," could "have a significant effect on climate." By the end of the 20th century, proposals to arrest climate change by limiting the use of fossil fuels had become a prominent public policy issue. Demands for taking steps to stabilize concentrations of carbon dioxide in the atmosphere, combined with the need for plants to generate sufficient electricity to meet projected energy requirements, gave new impetus to reconsidering nuclear power.

In 2000, a group of analysts from various fields of expertise argued in an article in Science magazine that "nuclear power can play a significant role in mitigating climate change." Their position received strong support in an interdisciplinary study conducted at MIT and published in 2003. The report, entitled, "The Future of Nuclear Power," pointed out that "over the next 50 years, unless patterns change dramatically, energy production and use will contribute to global warming through large-scale gas emissions—hundreds of billions of tonnes of carbon in the form of carbon dioxide." It concluded that the "nuclear option should be retained precisely because it is an important carbon-free source of power." It urged that steps be taken to expand knowledge of safety issues throughout the nuclear fuel cycle, to improve international safeguards, and to resolve the problem of waste disposal. The MIT study also called for tax credits and other financial incentives to encourage the construction of nuclear plants and other carbon-free sources of energy.

The capital costs of building a nuclear plant were widely viewed as the major deterrent to the growth of the industry. Despite the need for power, the fear of global warming, and growing public support for nuclear power, the Financial Times reported in 2005 that "investing the billions of dollars needed to construct new reactors remains an enormous gamble." That same year, Congress passed and President Bush signed an energy bill that sought to ease the financial burdens on utilities that built new nuclear plants with loan guarantees and other subsidies. The financial incentives were intended to encourage what was often referred to as the "nuclear renaissance." The revival of the nuclear option proceeded steadily but not without considerable uncertainty. In 2006, National Geographic ran an article entitled, "It's Scary. It's Expensive. It Could Save the Earth." It began by asking the question, "Nukes Again?" Its answer was equivocal: "Maybe." By June 30, 2009, the NRC had received 18 combined operating license applications for 28 new nuclear plants. However, only a few of the companies who submitted applications planned to start construction as soon as they received NRC approval.

Regardless of the extent of the nuclear renaissance or the pace at which it proceeds, the nuclear industry, the NRC, and other stakeholders should be keenly aware of the history of the first nuclear boom of the 1960s and mindful of the lessons to be learned from the experiences that followed, especially the need for conservative design, scrupulous operation, and careful regulation.

NUREG/BR-0175, Rev. 2
October 2010